# STUDENT PROBLEM MANUAL

for use with

---

# PRINCIPLES AND APPLICATIONS OF ELECTRICAL ENGINEERING

---

Second Edition

## Giorgio Rizzoni

Prepared by

### Bahman Samimy
*Ohio State University*

**IRWIN**

Chicago • Bogotá • Boston • Buenos Aires • Caracas
London • Madrid • Mexico City • Sydney • Toronto

ISBN 0–256–22498–6

1 2 3 4 5 6 7 8 9 0 WCB 3 2 1 0 9 8 7 6

## Introduction to the EIT Exam

Each of the 50 states regulates the engineering profession by requiring individuals who intend to practice the profession to become registered professional engineers (PE). To become a professional engineer, it is necessary to satisfy four requirements. The first is the completion of a B.S. degree in engineering from an accredited college or university (although it is theoretically possible to be registered without having completed a degree). The second is the successful completion fo the Engineer-in-Training (Engineering Fundamentals) Examination. This is an eight-hour exam that covers general engineering undergraduate education. The third requirement is two to four years of engineering experience after passing the Engineering-in-Training (EIT) exam. Finally, the fourth requirement is successfull completion of the Principles and Practice of Engineering or Professional Engineering (PE) Examination.

The EIT exam is a two-part national examination given twice a year (in April and October). The exam is divided into two four-hour sessions. The morning session consists of 140 multiple choice questions (five possible answers are given); the afternoon session consists of 70 questions. The exam is prepared by the National Council of Examiners of Engineering and Surveying.

One aim of this book is to assist you in preparing for one part of the EIT exam consists of a total of 14 questions in the morning session and 10 questions in the afternoon session. The examination topics for the electrical circuits part are the following:

      DC Circuits

      AC Circuits

      Three-Phase Circuits

      Capacitance and Inductance

      Transients

      Diode-Applications

      Operational Amplifiers (Ideal)

      Electric and Magnetic Fields

      Electric Machinery

Solved sample problems in these areas are given at the ends of many chapters, together with a summary of useful examples you may refer to in the text; all topics are eventually covered. You will find these end-of-chapter reviews invaluable in preparing for the Electrical circuits part of the EIT exam.

The detailed topics covered in the electrical circuits part of the EIT exam are listed below, along with the relevant chapter in the text.

      DC Circuits

            Relationship between current and charge (Chapter 2)

            Current and voltage laws (Chapter 2)

            Power and energy (Chapter 2)

            Equivalent series and parallel elements (Chapter 2)

Maximum power transfer (Chapter 3)

Thévenin and Norton theorems (Chapter 3)

AC Circuits

Algebra of complex numbers (Appendix A)

Phasor representation (Chapter 4)

Concept of time domain and frequency domain (Chapters 4 and 6)

Impedance concepts (Chapter 4)

Complex power, apparent power, and power factor (Chapter 5)

Ideal tranformers (Chapter 5)

Three-Phase Circuits

Computation of Power (Chapter 5)

Line and phase currents for delta and wye circuits (Chapter 5)

Capacitance and Inductance

Energy storage (Chapter 4)

$RL$ and $RC$ (first-order) transients (Chapter 6)

Transients (Chapter 6)

Diode Applications (Chapter 7)

Operational Amplifiers (Ideal) (Chapter 11)

Electric and Magnetic Fields

Work done in moving a charge in an electric field (Chapter 2)

Relationship between voltage and work (Chapter 15)

Faraday's and Ampère's laws (Chapter 15)

Magnetic reluctance concepts (Chapter 15)

Electric Machinery

AC and DC motor fundamentals (Chapter 16, 17)

# Section 1: Fundamentals of Electric Circuits

**1.1** A lightning bolt carries a current of 10,000 A and lasts for 100μsec.

    a) How much charge is exchanged between the clouds and the ground?

    b) If the charge differential between the clouds and the ground is caused by charge deficient raindrops, determine the number of raindrops must fall to create the charge separation which was neutralized by the lightning of part a). Each raindrop has a charge deficiency of 20 electronic charges.

**Solution:**

a) $i = \dfrac{dQ}{dt} \dfrac{C}{s}$

$$Q = \int_{0}^{10^{-4}} 10000A \; dt$$

$$\boxed{Q = 1 \; C}$$

b) $Q = n \; x \left(\dfrac{\#charges}{drop}\right) \times \left(\dfrac{Coulomb}{charge}\right) = 1 \; C$

$$n = \frac{1}{(-20)(-1.6 \times 10^{-19})}$$

$$\boxed{n = 3.12 \times 10^{17} \; \text{rain drops}}$$

**1.2** The battery-charging scheme shown in Figure P2.3 is an example of a constant current charge cycle. The voltage at the battery's terminals is created by the charger to be such that the current into the battery is held constant at 40mA. The battery is charged for 6 hours. Find

      a) the total charge transferred to the battery;

      b) the energy transferred to the battery during the 6 hours charge cycle.

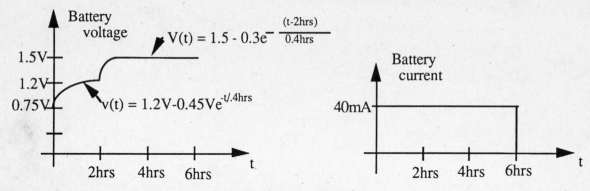

**Figure P2.3**

## Solution:

a) To find the charge delivered to the battery during the charge cycle we examine the charge-current relationship:

$$i = \frac{dq}{dt} \quad \text{or} \quad dq = i\,dt$$

thus

$$Q = \int_{t_o}^{t_1} i(t)dt$$

$$Q = \int_{o}^{6\ hrs} 40 \times 10^{-3}dt = \int_{o}^{21600\ s} 0.04\ dt = 21600 \times 0.04$$

$$\boxed{Q = 864\ C}$$

b) To find the energy transferred to the battery, we examine the energy relationship

$$\frac{dw}{dt} = P \qquad dw = p(t)\,dt$$

$$w = \int_{t_o}^{t_1} p(t)dt = \int_{t0}^{t_1} v(t)i(t)dt$$

$p(t) = v(t)i(t)$

$= 0.048 - 0.018\ e^{-t/0.4\ hrs}, \ 0 < t < 2\ hrs$

or $0.06 - 0.012e^{-(t - 2\ hrs)/0.4\ hrs}, \ 2hrs < t < 6hrs$

$$\therefore \ w = \int_{o}^{2\ hrs} (0.048 - 0.018\ e^{-t/0.4\ hrs})dt$$

$$+ \int_{2\ hrs}^{6hrs} (0.06 - 0.012\ e^{-(t-2\ hrs)/0.4\ hrs})\ dt$$

$$= [0.048t - 0.018(-1440)e^{-t/1440}]\Big|_{0}^{7200\ s}$$

$$+ [0.06t - (0.012)(-1440)\ e^{-t/1440}]\Big|_{0}^{14400\ s}$$

$$= 0.048\ (7200) + 25.92\ (0.006738 - 1) +$$

$$+ 0.06\ (14400) + 17.28\ (-1)$$

$$= 345.6 - 25.75 + 432 - 17.28$$

$$\boxed{w = 734.57\ J}$$

**1.3** The battery charging scheme shown in Figure P2.4 is called a " taper current charge cycle." The current starts off at its highest level and then decreases with time for the entire charge cycle as shown.    Find:

a) the total charge transferred to the battery;

b) the energy transferred to the battery in the 12 hour charge cycle.

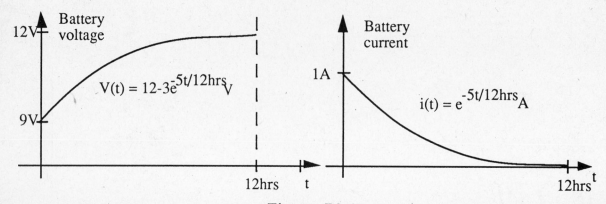

**Figure  P2.4**

1.4

## Solution:

a) To find the charge delivered to the battery during the charge cycle we examine the charge-current relationship:

$$i = \frac{dq}{dt} \quad \text{or} \quad dq = idt$$

thus

$$Q = \int_{t_o}^{t_1} i(t)dt$$

$$Q = \int_{o}^{12 \text{ hrs}} e^{-5t/12 \text{ hrs}} dt$$

$$= \int_{o}^{43,200 \text{ s}} e^{-5t/43,200 \text{ s}} dt$$

$$= -\frac{43200}{5} e^{-5t/43200 \text{ s}} \Big|_{0}^{43200}$$

$$= -\frac{43200}{5}(e^{-5} - e^{0})$$

$$= -\frac{43200}{5}(0.006737 - 1)$$

$$\boxed{Q = 8582 \text{ C}}$$

b)To find the energy transferred to the battery, we examine the energy relationship

$$\frac{dw}{dt} = P \quad dw = p(t) \, dt$$

$$w = \int_{t_o}^{t_1} p(t) \, dt = \int_{t_0}^{t_1} v(t)i(t)dt$$

$$p(t) = v(t)i(t)$$

$$= 12e^{-5t/12 \text{ hrs}} - 3e^{-10t/12 \text{ hrs}}$$

$$w = \int_{0}^{43200 \text{ s}} 12e^{-5t/43200s} - 3e^{-10t/43200s} \, dt$$

$$= \{-\frac{43200}{5}(12)e^{-5t/43200}$$

$$- \frac{43200}{10}(-3)e^{-10t/43200}\}\Big|_{0}^{43200}$$

$$= -103,680 (e^{-5} - e^{0}) + 12,960(e^{-10} - e^{0})$$

$$= -103680(-0.99326) + 12960(-0.999955)$$

$$\boxed{w = 90.0 \text{ kJ}}$$

**1.4** Apply KVL to find the voltage $v$ in the circuit of Figure P2.8.

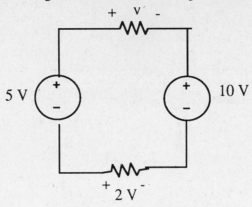

**Figure P2.8**

**Solution:**

Applying KVL:

$$-5 + v + 10 - 2 = 0$$
$$v = -3 \text{ V}$$

1.6

**1.5**   Determine which elements in the circuit of Figure P2.13 are supplying power and which are dissipating power.  Also determine the amount of power dissipated and supplied.

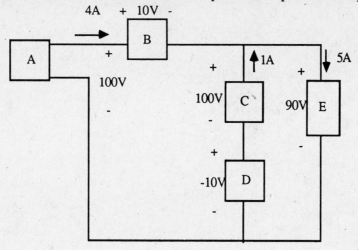

**Figure  P2.13**

**Solution:**
Element A:

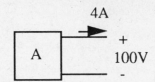

$P = vi = 100(-4) = -400$ W (supplied)

Element B:

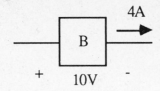

$P = vi = (10)(4) = 40$ W (dissipated)

Element C:

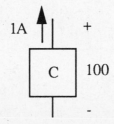

$P = vi = (100)(-1) = -100$ W  (supplied)

Element D:

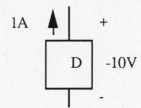

$P = vi = (-10)(-1) = 10$ W (dissipated)

Element E:

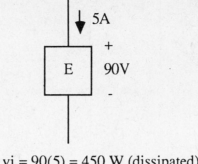

$$P = vi = 90(5) = 450 \text{ W (dissipated)}$$

---

**1.6**  Determine which elements in the circuit of Figure P2.14 are supplying power and which is dissipating power.  Also determine the amount of power dissipated and supplied.

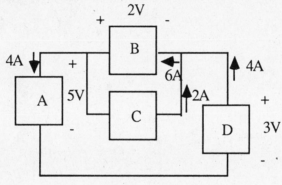

**Figure  P2.14**

1.9

**Solution:**

Element A:

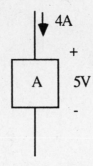

$$P = vi = (5)(4) = 20 \text{ W (dissipated)}$$

Element B:

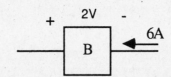

$$P = v_i = (2)(-6) = -12 \text{ W (supplied)}$$

Element C:

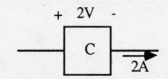

$$P = v_i = (2)(2) = 4 \text{ W (dissipated)}$$

Element D:

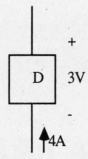

$$P = v_i = 3(-4) = -12 \text{ W (supplied)}$$

**1.7**    Find the current i and the power delivered to the resistor in the circuit of Figure P2.15.

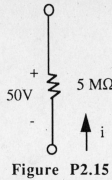

**Figure   P2.15**

**Solution:**

Applying Ohm's Law:

$$i = \frac{-50}{5 \times 10^6} = -10^{-5} \text{ A} = -10 \, \mu\text{A}$$

$$P = vi = 50 \times 10^{-5} = 0.5 \text{ mW}$$

**1.8**   Use Kirchhoff's voltage law to determine the voltage across <u>each</u> of the resistors in the circuit of Figure P2.20.

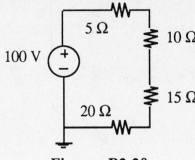

**Figure  P2.20**

**Solution:**

Applying KVL:

$$-100 + 5i + 10i + 15i + 20i = 0$$

Therefore

$$i = 2 \text{ A}$$

Thus, we can apply Ohm's law to find the voltage across each resistor.:

$$v_{5\,\Omega} = 10 \text{ V}, \ v_{10\,\Omega} = 20 \text{ V}, \ v_{15\,\Omega} = 30 \text{ V}$$

and

$$v_{20\,\Omega} = 40 \text{ V}$$

This problem can also be solved using the voltage divider rule.

**1.9**   Use Ohm's law and KVL to determine the voltage $V_1$ in the circuit of Figure P2.21.

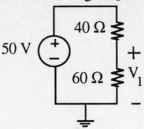

**Figure  P2.21**

**Solution:**

Applying KVL:

$$-50 + 40i + 60i = 0$$

therefore

$$i = 0.5 \text{ A}$$

Now we can apply Ohm's law to find the voltage $V_1$.

$$V_1 = 60i = 30 \text{ V}$$

**1.10** Find the equivalent resistance seen by the source and the current $i$. in the circuit of Figure P2.28.

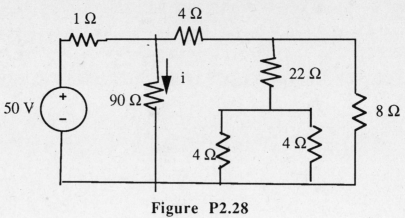

**Figure  P2.28**

## Solution:

Step1: $(4 \parallel 4) + 22 = 24 \Omega$

Step 2: $24 \parallel 8 = 6$

Therefore, the equivalent circuit is as shown below:

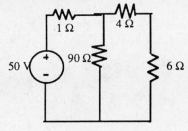

Further, $(4 + 6) \parallel 90 = 9 \Omega$

The new equivalent circuit is shown below.

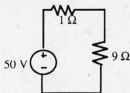

Thus, $\boxed{R_{Total} = 10 \; \Omega}$

We can now find the current i by the current divider rule as follows:

$$i = (\frac{10}{10 + 90}) \, (5) = 0.5 \; A$$

**1.11** In the circuit of Figure P2.30, find the equivalent resistance, where $R_1 = 5\Omega$, $R_2 = R_6 = 1K\Omega$, $R_3 = R_4 = 100\Omega$, $R_5 = 9.1\Omega$.

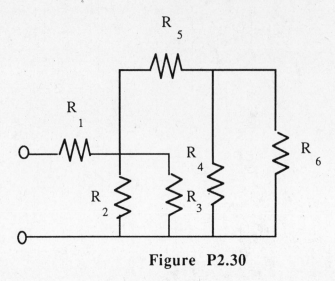

**Figure P2.30**

**Solution:**

Step1: $(R_6 \| R_4) + R_5 = (1{,}000\ \Omega \| 100\ \Omega) + 9.1\ \Omega \approx 100\ \Omega$, resulting in the circuit shown below.

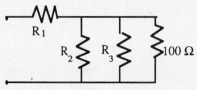

Therefore, the equivalent resistance is

$$R_{eq} = (100\ \| R_3 \| R_2) + R_1$$
$$= (100 \| 100 \| 1000) + 5 = 52.6\ \Omega$$

**1.12** The circuit of Figure P2.34, which contains a dependent source, is a model for a *transistor*. Find the output voltage if $v_{in} = 0.5V$.

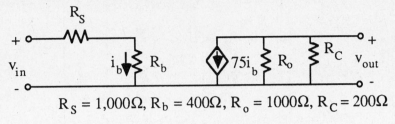

$$R_S = 1,000\Omega, R_b = 400\Omega, R_o = 1000\Omega, R_C = 200\Omega$$

**Figure P2.34**

**Solution:**

Applying KVL:

$$v_{in} = R_s i_b + R_b i_b = (R_s + R_b) i_b$$

Therefore, $i_b = \dfrac{1}{2800} A$

$$v_{out} = -75 i_b (R_C \parallel R_o) = -4.46 \text{ V}$$

**1.13** Find the current, $i_O$, in the circuit of Figure P2.35. Assume that $v_S = 0.2V$, $R_S = 100\Omega$, R
10K$\Omega$,

$R_b = 500\Omega$, $R_o = 5K\Omega$, $R_D = 50\Omega$.

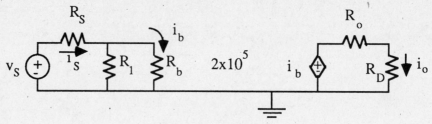

**Figure P2.35**

**Solution:**

Using the current divider method:

$$i_b = \frac{R_1}{R_1+R_b} i_S$$

$$i_S = \frac{v_S}{R_S+R_1\|R_b}$$

$$i_o = \frac{2\times10^5 i_b}{R_o+R_D} = \frac{2\times10^5}{R_o+R_D} \times \frac{R_1}{R_1+R_b} \times \frac{v_S}{R_S+R_1\|R_b}$$

$$i_o = \frac{2\times10^5\times10k\times0.2}{5050\times10500(100+10k\|500)} =$$

$$= 13.09 \ mA$$

**1.14** With reference to Figure 2.42,

a. find a general expression for the power internally dissipated by the wattmeter as a function of the resistors R1 and R2, assuming that the voltmeter has an internal resistance $r_V$, and the ammeter has an internal resistance $r_A$.

b. Compute the percent error in the power measurement for this wattmeter if the $R_1 = 50\Omega$, $R_2 = 10k\Omega$, $r_A = 1.2k\Omega$ and $r_V = 1M\Omega$.

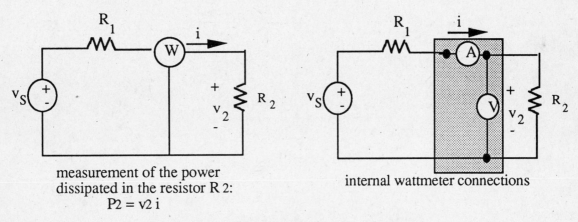

measurement of the power
dissipated in the resistor R 2:
$P2 = v2\ i$

internal wattmeter connections

**Figure 2.42 Measurement of power**

**Solution:**

The equivalent circuit is as shown below.

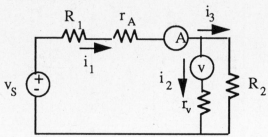

Using KVL:

$$v_s = R_1 i_1 + r_A i_1 + (r_V \parallel R_2) i_1$$

$$v_s = (R_1 + r_A + \frac{r_V R_2}{r_V + R_2}) i_1$$

Therefore,

$$i_1 = \frac{v_s(r_V + R_2)}{R_1 r_V + R_1 R_2 + r_A r_V + r_A R_2 + r_V R_2}$$

Also, by current division:

$$i_2 = \frac{R_2}{R_2 + r_V} i_1$$

$$P_{meter} = (i_1)^2 r_A + (i_2)^2 r_V$$

$$P_{meter} =$$
$$= (\frac{v_s(r_V + R_2)}{R_1 r_V + R_1 R_2 + r_A r_V + r_A R_2 + r_V R_2})^2 r_A +$$
$$(\frac{v_s R_2}{R_1 r_V + R_1 R_2 + r_A r_V + r_A R_2 + r_V R_2}))^2 r_V$$

b) The true power dissipated without the meter in the circuit ($P_{no\ loss}$) is:

$$P_{no\ loss} = (\frac{v_s}{R_1 + R_2})^2 R_2 = \frac{10000}{(10050)^2} (v_s)^2$$

The actual power measured by the meter ($P_{loss}$) is:

$$P_{loss} = (i_3)^2 R_2$$

where $i_3 = \frac{r_V}{R_2 + r_V} i_1$

therefore,

$$P_{loss} = \frac{10000}{(11262)^2} (v_s)^2$$

Thus,

$$\%Error = (1 - \frac{P_{loss}}{P_{no\ loss}}) \times 100 = 20.4\%$$

**1.15** For the wattmeter of Figure 2.42, generate a graph of the percent error in the power measurement as a function of $R_2$, assuming $R_2$ can vary between 1 and 100 k$\Omega$. Can you draw a general conclusion with regard to the accuracy of power measurements?

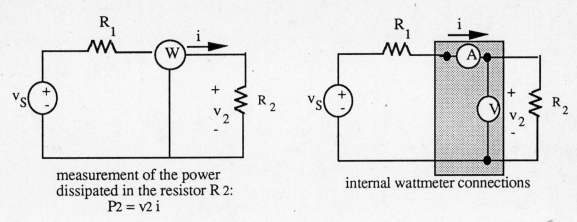

measurement of the power
dissipated in the resistor R 2:
$$P2 = v2\ i$$

internal wattmeter connections

**Figure 2.42 Measurement of power**

## Solution:

A short BASIC program may be used to generate a few data points for the curve. Using the results of Problem 2.53, we have:

```
10000 LPRINT "R2",,"% Error"

10010 R1=50

10020 R2=1

10030 RA=1200

10040 RV=1000000!

10050 I1=(RV+R2)/(R1*RV+R1*R2+RA*RV+RA*R2+RV*R2)

10060 I3=RV*I1/(R2+RV)

10070 PLOSS=I3*I3*R2

10080 PNOLOSS=R2/(R1+R2)/(R1+R2)

10090 PCTERROR=(1-PLOSS/PNOLOSS)*100

10100 LPRINT R2,,PCTERROR

10110 R2=10*R2

10120 IF R2<1000000! THEN 10050

10130 END
```

The output of this program is given below:

| R2 | % Error |
|---|---|
| 1 | 99.8338 |
| 10 | 99.77324 |
| 100 | 98.76566 |
| 1000 | 78.2464 |
| 10000 | 20.37261 |
| 100000 | 2.596974 |

Clearly, the larger the value of R2, the smaller the error.

## Section 2: Resistive Network Analysis

**2.1** Find the currents $i_1$ and $i_2$ for the circuit of Figure 3.18 using node voltage analysis.

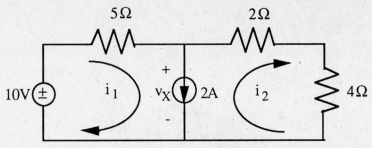

**Figure 3.18**

**Solution:**

Applying nodal analysis;

$$\frac{v_x - 10}{5} + \frac{v_x}{6} + 2 = 0$$

solving the equation, we obtain

$$v_x = 0 \text{ V}$$

Therefore,

$$i_1 = \frac{10 - v_x}{5} = \frac{10 - 0}{5} = 2 \text{ A}$$

$$i_2 = \frac{v_x}{2 + 4} = \frac{0}{6} = 0 \text{ A}$$

**2.2**  Find the currents $i_1$, $i_2$ and $i_3$ for the circuit of Figure 3.19 using node voltage analysis.

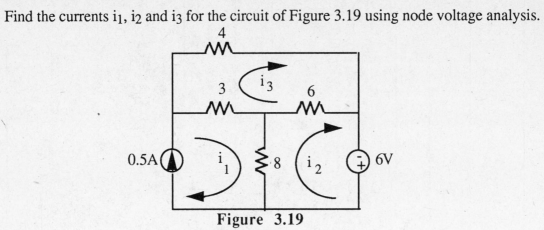

Figure 3.19

## Solution:

Using node voltage analysis at the three labeled nodes, we have

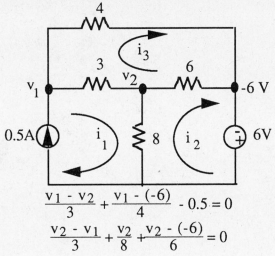

$$\frac{v_1 - v_2}{3} + \frac{v_1 - (-6)}{4} - 0.5 = 0$$

$$\frac{v_2 - v_1}{3} + \frac{v_2}{8} + \frac{v_2 - (-6)}{6} = 0$$

Rearranging the equations,

$$7v_1 - 4v_2 = -12$$

$$-8v_1 + 15v_2 = -24$$

Solving the two equations, we have

$$v_1 = -3.781 \text{ V and } v_2 = -3.616 \text{ V}$$

therefore,

$$i_1 = 0.5 \text{ A}$$

$$i_2 - i_3 = \frac{v_2 - (-6)}{6} = \frac{-3.616 + 6}{6} = 0.397 \text{ A}$$

$$i_3 = \frac{v_1 - (-6)}{4} = \frac{-3.781 + 6}{4} = 0.555 \text{ A}$$

$$i_2 = 0.397 + 0.555 = 0.952 \text{ A}$$

**2.3**  Find the voltages $V_1$ and $V_2$ for the circuit of Figure 3.16 using node voltage analysis.

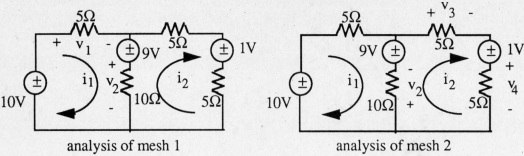

analysis of mesh 1          analysis of mesh 2

**Figure 3.16**

**Solution:**

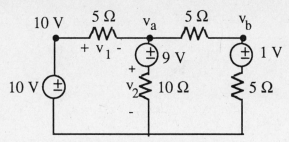

We need to find $v_a$ first. Then, using the following equations:

$$v_1 = 10 - v_a$$

$$v_2 = v_a - 9$$

we can find $v_1$ and $v_2$. Using nodal analysis, we write

$$\frac{v_a - 10}{5} + \frac{v_a - 9}{10} + \frac{v_a - v_b}{5} = 0$$

$$\frac{v_b - v_a}{5} + \frac{v_b - 1}{5} = 0$$

Rearranging the equations,

$$5v_a - 2v_b = 29$$

$$-v_a + 2v_b = 1$$

we can solve for $v_a$, $v_1$ and $v_2$:

$$v_a = 7.5 \text{ V}$$

$$v_1 = 2.5 \text{ V and } v_2 = -1.5 \text{ V}$$

---

**2.4** Find the unknown current $i_x$ for the circuit of Figure 3.21 using node voltage analysis.

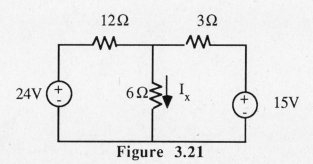

**Figure 3.21**

**Solution:**

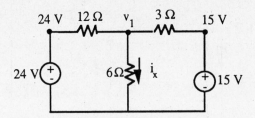

Using nodal analysis:

$$\frac{v_1 - 24}{12} + \frac{v_1}{6} + \frac{v_1 - 15}{3} = 0$$

$$v_1 = 12 \text{ V}$$

Therefore,

$$i_x = \frac{v_1}{6} = \frac{12}{6} = 2 \text{ A}$$

---

**2.5** Find the unknown voltage $V_x$ for the circuit of figure 3.20 using node voltage analysis.

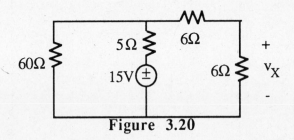

**Figure 3.20**

**Solution:**

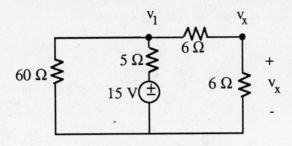

Using nodal analysis:

$$\frac{v_1}{60} + \frac{v_1 - 15}{5} + \frac{v_1 - v_x}{6} = 0$$

$$\frac{v_x - v_1}{6} + \frac{v_x}{6} = 0$$

Rearranging the equations,

$$23 v_1 - 10 v_x = 180$$

$$-v_1 + 2 v_x = 0$$

Solving for the two unknowns,

$$v_1 = 10 \text{ V and } v_x = 5 \text{ V}$$

---

**2.6** Using mesh current analysis, find the currents $I_1$ and $I_2$ and the voltage across the 20 Ω resistor for the circuit of Figure P3.6.

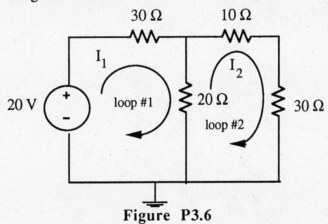

**Figure P3.6**

**Solution:**

Mesh #1:    $20 = 30 I_1 + 20 (I_1 - I_2)$

Mesh #2:   $20 (I_2 - I_1) + 10 I_2 + 30 I_2 = 0$

Simplifying the above equations:

$$50 I_1 - 20 I_2 = 20$$
$$-20 I_1 + 60 I_2 = 0$$

Therefore,

$$I_1 = 0.462 \text{ A and } I_2 = 0.154 \text{ A}$$

The voltage across the 20-$\Omega$ resistor is then

$$20 (I_1 - I_2) = 6.15 \text{ V} \quad (+ \text{ ref. at top})$$

---

**2.7**    Find the voltage across the 1k$\Omega$ resistor in Figure 3.9 using mesh current analysis.

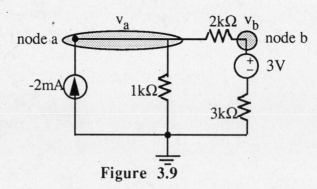

**Figure 3.9**

## Solution:

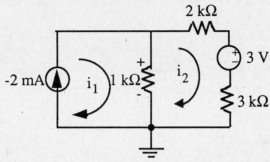

Using units of kΩ and mA, we write the equations:

Mesh #1      $i_1 = -2$

Mesh #2      $(1 + 2 + 3)\, i_2 - (1)\, i_1 = -3$

Solving for $i_2$,

$$i_2 = \frac{-5}{6}\, mA$$

$$v_{1\,k\Omega} = R_{1\,k\Omega}(i_1 - i_2)$$
$$v_{1\,k\Omega} = 1(-2 - (-\frac{5}{6})) = -1.67\ V$$

---

**2.8**    Find the voltage across the 10 Ω resistor using mesh current analysis.

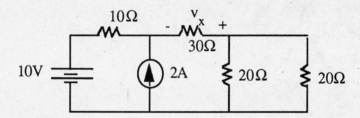

**Solution:**

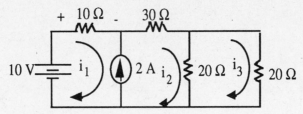

Current source constraint: $i_2 - i_1 = 2$

Mesh #1 and #2 combined:

$$10i_1 + 30i_2 + 20(i_2 - i_3) = 10$$

Mesh #3: $\quad 20i_3 + 20(i_3 - i_2) = 0$

Rearranging the equations:

$$-i_1 + i_2 = 2$$
$$i_1 + 5i_2 - 2i_3 = 1$$
$$i_2 - 2i_3 = 0$$

Therefore,

$$i_1 = -1.4 \text{ A} \quad i_2 = 0.6 \text{ A} \quad i_3 = 0.3 \text{ A}$$
$$v_{10\,\Omega} = 10(i_1) = -14 \text{ V}$$

---

**2.9** Find the voltage across the 10 kΩ resistor in Figure 3.4 using mesh current analysis.

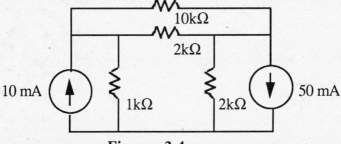

**Figure 3.4**

**Solution:**

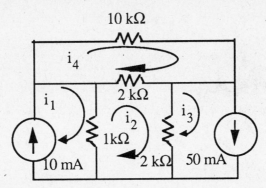

10 kΩ

$i_4$

$i_1$

2 kΩ

$i_2$

$i_3$

1kΩ

10 mA

2 kΩ

50 mA

From the figure, we can immediately see that

$i_1 = 10$ mA and $i_3 = 50$ mA. Using mesh analysis, with resistances in kΩ and currents in mA,

Mesh #2

$$(-1) i_1 + (1 + 2 + 2) i_2 - 2 i_3 - 2 i_4 = 0$$

Mesh #4

$$(-2) i_2 + (2 + 10) i_4 = 0$$

Solving for the unknowns,

$i_2 = 23.571$ mA and $i_4 = 3.929$ mA.

Therefore,

$$v_{10 kΩ} = 10 i_4 = 39.29 \text{ V}$$

---

**2.10** Find the voltage across the $R_2$ resistor in Figure 3.6 using mesh current analysis. Assume $R_1 = 1K\Omega$, $R_2 = 500 \ \Omega$, $R_3 = 2.2 \ k\Omega$, $R_4 = 4.7 \ k\Omega$, $i_a = 1mA$, $i_b = 2mA$.

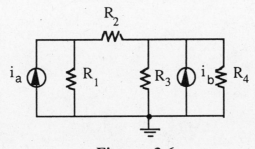

$R_2$

$i_a$   $R_1$   $R_3$   $i_b$   $R_4$

**Figure 3.6**

**Solution:**

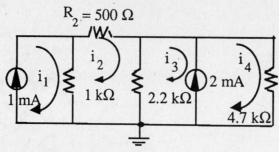

From the figure, we can immediately see that

$$i_1 = 1 \text{ mA and } i_4 - i_3 = 2 \text{ mA.}$$

Using mesh analysis, with resistance in $k\Omega$ and currents in mA:

Meshes #3 and 4 combined

$$-1 \, i_1 + (1 + 0.5 + 2.2) \, i_2 - 2.2 \, i_3 = 0$$

Mesh #2 and 4

$$-2.2 \, i_2 + 2.2 \, i_3 + 4.7 \, i_4 = 0$$

Solving for the unknowns,

$i_2 = -0.666$ mA, $i_3 = -1.575$ mA and

$$i_4 = 0.425 \text{ mA}$$

Therefore, $v_{R2} = 500 \, i_2 = -0.333$ V

---

**2.11** Using mesh current analysis, solve for all unknown currents and voltages in Figure 3.36 given that $R_L = 10 \ \Omega$.

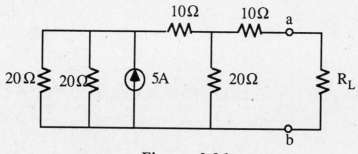

**Figure 3.36**

**Solution:**

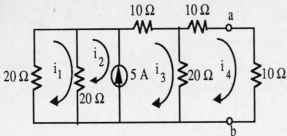

By inspection,   $i_3 - i_2 = 5$ A

Also, using mesh analysis:

Mesh #1

$$(20 + 20)\, i_1 - 20\, i_2 = 0$$

Meshes #2 and #3

$$-20\, i_1 + 20\, i_2 + (10 + 20)\, i_3 - 20\, i_4 = 0$$

Mesh #4

$$(20 + 10 + 10)\, i_4 - 20\, i_3 = 0$$

Rearranging and simplifying the equations,

$$-i_2 + i_3 = 5$$
$$2\, i_1 - i_2 = 0$$
$$-2\, i_1 + 2\, i_2 + 3\, i_3 - 2\, i_4 = 0$$
$$-i_3 + 2\, i_4 = 0$$

Therefore,

$$i_1 = -1.667 \text{ A} \quad i_2 = -3.33 \text{ A}$$
$$i_3 = 1.667 \text{ A} \quad i_4 = 0.833 \text{ A}$$

The unknown voltages are shown on the redrawn circuit.

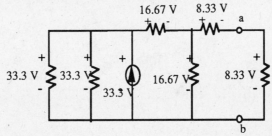

**2.12** Using mesh current analysis, solve for all unknown currents and voltages in Figure 3.38 given that $R_L = 2\ \Omega$.

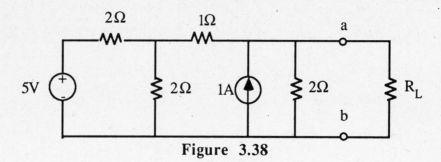

**Figure 3.38**

**Solution:**

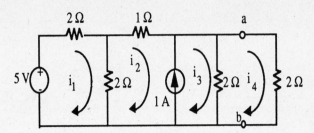

By inspection, $i_3 - i_2 = 1$ A

Also, for mesh #1     $(2 + 2) i_1 - 2 i_2 = 5$

Meshes #2 and #3

$$-2 i_1 + (2 + 1) i_2 + 2 i_3 - 2 i_4 = 0$$

Mesh #4     $(2 + 2) i_4 - 2 i_3 = 0$

Rearranging and simplifying the equations,

$$-i_2 + i_3 = 1$$
$$4 i_1 - 2 i_2 = 5$$
$$-2 i_1 + 3 i_2 + 2 i_3 - 2 i_4 = 0$$
$$-i_3 + 2 i_4 = 0$$

Therefore,

$$i_1 = 1.5 \text{ A} \qquad i_2 = 0.5 \text{ A}$$
$$i_3 = 1.5 \text{ A} \qquad i_4 = 0.75 \text{ A}$$

The unknown voltages are shown on the redrawn circuit.

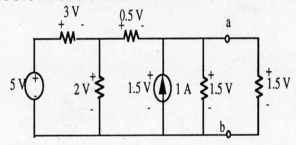

**2.13** Using mesh current analysis, solve for all unknown currents and voltages of the circuit given in Figure 3.41. Assume that $R_L$= 5 kΩ.

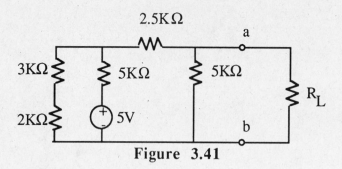

**Figure 3.41**

**Solution:**

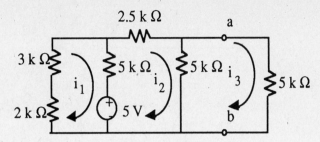

Using mesh analysis, with resistance in kΩ and current in mA:

Mesh #1

$$(3 + 2 + 5) i_1 - 5 i_2 = -5$$

Mesh #2

$$-5 i_1 + (5 + 2.5 + 5) i_2 - 5 i_3 = 5$$

Mesh #3

$$(5 + 5) i_3 - 5 i_2 = 0$$

Rearranging and simplifying the equations,

$$2 i_1 - i_2 = -1$$
$$-i_1 + 2.5 i_2 - i_3 = 1$$
$$-i_2 + 2 i_3 = 0$$

Therefore,

$$i_1 = -\frac{1}{3} \text{ mA} \qquad i_2 = \frac{1}{3} \text{ mA}$$
$$i_3 = \frac{1}{6} \text{ mA}$$

The unknown voltages are shown on the redrawn circuit.

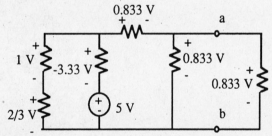

**2.14** Find the Thevenin equivalent voltage for the circuit of Figure 3.41.

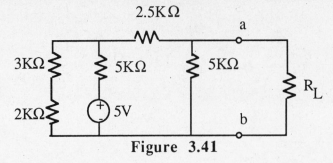

**Figure 3.41**

**Solution:**

After removing the load, and using mesh analysis, let the current in the left mesh be $i_1$ (clockwise), and the current in the right mesh be $i_2$ (clockwise). Then,

$$10 \times 10^3 \, i_1 - 5 \times 10^3 \, i_2 = -5$$
$$-5 \times 10^3 \, i_1 + 12.5 \times 10^3 \, i_2 = 5$$

Solve for $i_2 = 0.25$ mA, and obtain

$$V_T = 5 \times 10^3 \times i_2 = 1.25 \text{ V}$$

---

**2.15** Find the Norton equivalent current for the circuit of Figure 3.41.

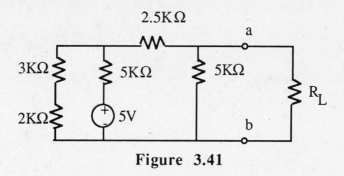

**Figure 3.41**

## Solution:

To find $i_N$, we use mesh analysis:

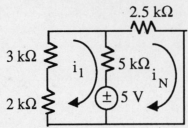

One 5 k-$\Omega$ resistor is omitted because the short circuit is connectd directly across it.

For mesh #1    $10i_1 - 5i_N = -5$,  and

for mesh #2    $-5i_1 + 7.5i_N = 5$

where resistances are in k$\Omega$ and currents are in mA.  Then, solving

$$i_N = 0.5 \text{ mA}$$

If one were also interested in the Thèvenin resistance, it could also computed as follows:

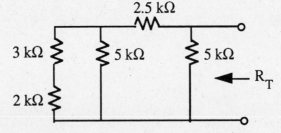

Therefore,

$$R_T = \{(3 + 2)\| 5 + 2.5\} \| 5 = 2.5 \text{ k}\Omega$$

**2.16** Find the Thevenin equivalent voltage for the circuit of Figure 3.42.

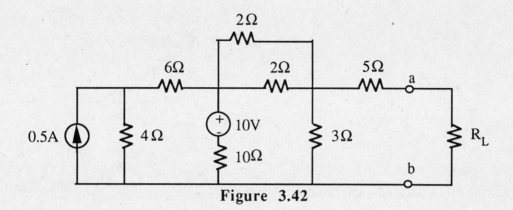

**Figure 3.42**

## Solution:

To find $v_T = v_{OC}$ we use the circuit shown.

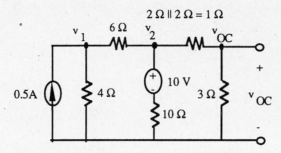

Note that the 5-$\Omega$ resistor is eliminated from the circuit since there is no current flowing through this resistor when the load is an open circuit. Applying node voltage analysis to find $v_{OC}$, we write the following equations:

$$(\frac{1}{4} + \frac{1}{6})v_1 - \frac{1}{6}v_2 = 0.5$$

$$-\frac{1}{6}v_1 + (\frac{1}{6} + 1 + \frac{1}{10})v_2 - v_{OC} = 1$$

$$-v_2 + (1 + \frac{1}{3})v_{OC} = 0$$

solving for $v_{OC}$,

$$v_{OC} = 2 \text{ V}$$

---

**2.17** Find the Norton equivalent current for the circuit of Figure 3.42.

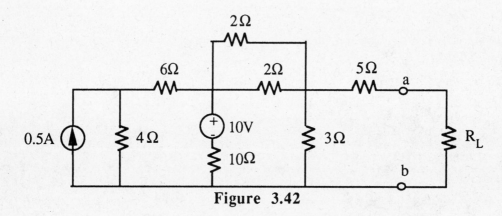

**Figure 3.42**

**Solution:**

The Norton current is shown below.

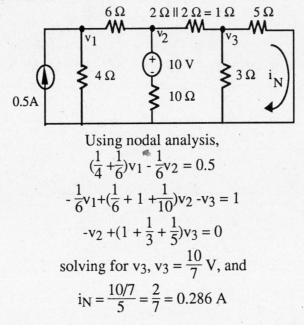

Using nodal analysis,

$$(\frac{1}{4} + \frac{1}{6})v_1 - \frac{1}{6}v_2 = 0.5$$

$$-\frac{1}{6}v_1 + (\frac{1}{6} + 1 + \frac{1}{10})v_2 - v_3 = 1$$

$$-v_2 + (1 + \frac{1}{3} + \frac{1}{5})v_3 = 0$$

solving for $v_3$, $v_3 = \frac{10}{7}$ V, and

$$i_N = \frac{10/7}{5} = \frac{2}{7} = 0.286 \text{ A}$$

---

**2.18** Find the Thevenin equivalent voltage for the circuit of Figure 3.43.

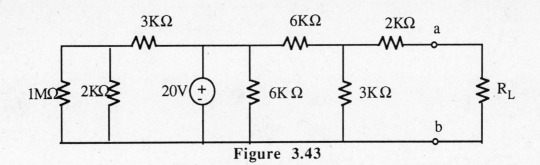

**Figure 3.43**

**Solution:**

The network to the left of the voltage source plays no part in the calculation of the Thèvenin voltage (why?). Similarly, the

6- kΩ resistor in parallel with the source has no effect. Using voltage division, we obtain

$$v_{OC} = \frac{3,000}{3,000+6,000} \times 20 = 6.67 \text{ V}$$

---

**2.19** Find the Norton equivalent current for the circuit of Figure 3.43.

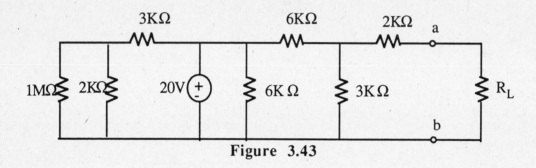

**Figure 3.43**

**Solution:**

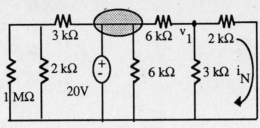

Applying nodal analysis, with resistance values in k$\Omega$,

$$(\tfrac{1}{3} + \tfrac{1}{2} + \tfrac{1}{6})v_1 - (\tfrac{1}{6})20 = 0$$

Solving for $v_1$, $v_1 = \dfrac{10}{3}$ V

Therefore, $i_N = \dfrac{v_1}{2000} = 1.667$ mA

---

**2.20** Find the Thevenin equivalent voltage for the circuit of Figure 3.44.

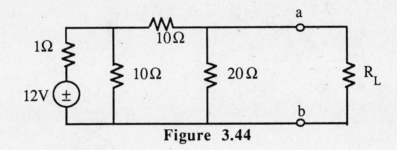

**Figure 3.44**

**Solution:**

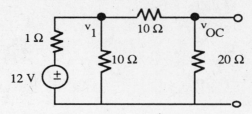

Using node voltage analysis,

$$(1 + \frac{1}{10} + \frac{1}{10})v_1 - \frac{1}{10}v_{OC} = 12$$

$$(\frac{1}{10} + \frac{1}{20})v_{OC} - \frac{1}{10}v_1 = 0$$

or

$$12v_1 - v_{OC} = 120$$

$$-2v_1 + 3v_{OC} = 0$$

Therefore,

$$\boxed{v_{OC} = 7.06 \text{ V}}$$

---

**2.21** Find the Norton equivalent current for the circuit of Figure 3.44.

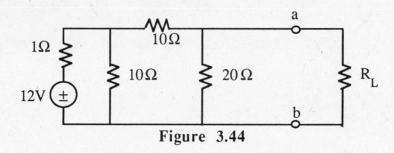

**Figure 3.44**

● **Solution:**

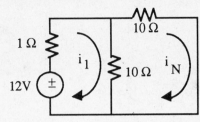

The 20-Ω resistor is omitted since the short circuit is directly across it. Using mesh analysis,

mesh #1 $(1+10)i_1 - 10i_N = 12$

mesh #2 $-10i_1 + (10+10)i_N = 0$

or

$$11i_1 - 10i_N = 12$$
$$-i_1 + 2i_N = 0$$

Thus,

$$i_N = 1 \text{ A}$$

---

**2.22** A resistive "pad", shown in Figure P3.20, is often used in order to make sources and loads which have different resistances appear to each other to be of compatible magnitude.

● If $R_S$, the internal source resistance is 600Ω, $R_1 = 600Ω$, $R_2 = 1.2Ω$, and $R_3 = 1KΩ$

    a) find the equivalent circuits seen by the source and load respectively.
    (Note that $R_2 \ll R_3 + R_L$).
    b)also find the Thevenin voltage that the load sees as a function of $V_S$

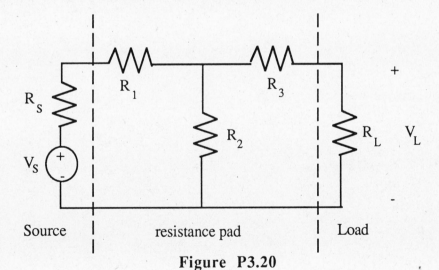

**Figure P3.20**

**Solution:**

$R_{eq}$ as seen by the source is

$$R_{eq} = R_1 + R_2 \| (R_3 + R_L)$$

But,

$$R_2 \ll R_3 + R_L$$

Therefore,

$$R_{eq} = R_1 + R_2$$

$$R_{eq} = 601.2 \ \Omega$$

$R_T$ as seen by the load is

$$R_T = R_3 + R_2 \| (R_S + R_1) = R_3 + R_2$$

$$\boxed{R_T = 1001.2 \ \Omega}$$

$$V_T = V_{OC} = \frac{R_2}{R_S + R_1 + R_2} V_S$$

$$\approx \frac{R_2}{R_S + R_1} V_S$$

$$\boxed{V_{OC} = 10^{-3} \ V_S}$$

**2.23** A system of equations can be expressed in matrix form for example:
$$[A][x] = [u]$$
and solved by taking the inverse of A and pre-multiplying through the equation.
$$[A]^{-1}[A][x] = [A]^{-1}[u]$$

$$[x] = [A]^{-1}[u]$$
Thus we can see that since we have a system of equations when we write node voltage and mesh current equations we can also write
$$[G][v_x] = [i]$$

$$[R][i_x] = [v]$$
Use the nodal analysis to find the matrices required to solve the circuit shown in Figure P3.22. Solve the circuit using computer aids to calculate the matrix inverses.

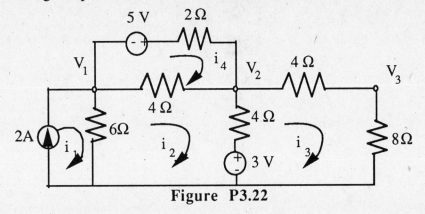

**Figure P3.22**

## Solution:

Using source transformations, the circuit simplifies to:

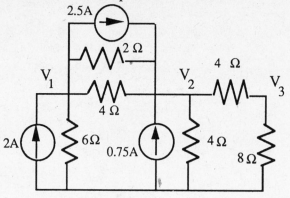

which yields

$$\begin{bmatrix} \frac{1}{6}+\frac{1}{4}+\frac{1}{2} & -(\frac{1}{2}+\frac{1}{4}) & 0 \\ -(\frac{1}{2}+\frac{1}{4}) & \frac{1}{4}+\frac{1}{2}+\frac{1}{4}+\frac{1}{4} & -\frac{1}{4} \\ 0 & -\frac{1}{4} & \frac{1}{4}+\frac{1}{8} \end{bmatrix} \begin{bmatrix} V_1 \\ V_2 \\ V_3 \end{bmatrix}$$

$$= \begin{bmatrix} 2-2.5 \\ 2.5+0.75 \\ 0 \end{bmatrix}$$

or

$$\begin{bmatrix} 11 & -9 & 0 \\ -3 & 5 & -1 \\ 0 & -2 & 3 \end{bmatrix} \begin{bmatrix} V_1 \\ V_2 \\ V_3 \end{bmatrix} = \begin{bmatrix} -6 \\ 13 \\ 0 \end{bmatrix}$$

This system of equations may be solved using a matrix algebra software package. The results are:

$$V_1 = 4.40 \text{ V}$$
$$V_2 = 6.05 \text{ V}$$
$$V_3 = 4.03 \text{ V}$$

---

**2.24** For the circuit of Figure P3.22 use the mesh current analysis to find the matrices required to solve the circuit and solve for the unknown currents.
[Hint: You may find it useful to use source transformations].

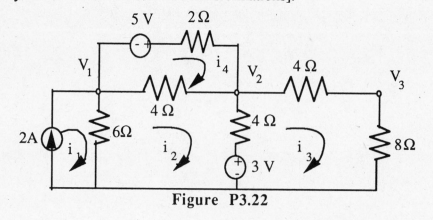

**Figure P3.22**

## Solution:

Using the hint, the circuit simplifies to:

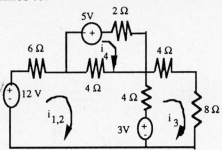

which yields

$$\begin{bmatrix} 6+4+4 & -4 & -4 \\ -4 & 4+4+8 & 0 \\ -4 & 0 & 4+2 \end{bmatrix} \begin{bmatrix} i_{1,2} \\ i_3 \\ i_4 \end{bmatrix}$$

$$= \begin{bmatrix} 12-3 \\ 3 \\ 5 \end{bmatrix}$$

The solutions are:

$$i_{1,2} = -1.55 \text{ A}$$

$$i_3 = -0.2 \text{ A}$$

$$i_4 = -7.475 \text{ A}$$

**2.25**  The voltage current relationship of a semiconductor diode may be approximated by the expression

$$i_D = I_{SAT} \left( \exp\left\{ \frac{V_D}{kT/q} \right\} - 1 \right)$$

where $I_{SAT} = 10^{-12}$ Amps

$\dfrac{KT}{q} = 0.0259$ volts at room temperature

and $V_D$ is the voltage across the diode.

a)  Given the circuit of Figure P3.34 use graphical analysis to find the diode current $i_0$ and the diode voltage $V_D$ if $R_T = 22\Omega$, and $V_T = 12V$.

b)  Write a FORTRAN or BASIC program which will find the diode voltage $V_D$ and the current $i_0$ using the flow chart given Figure P3.34.

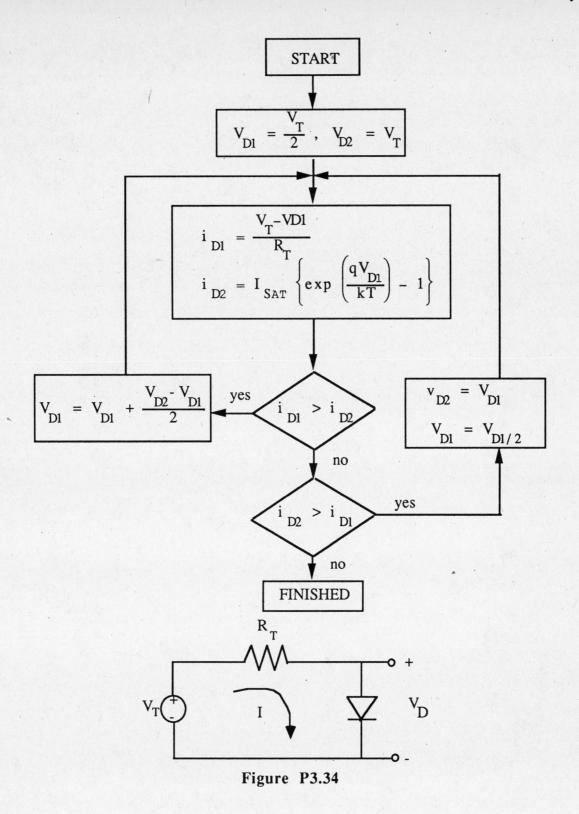

Figure P3.34

## Solution:

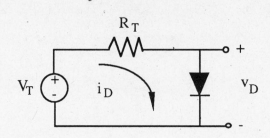

a)

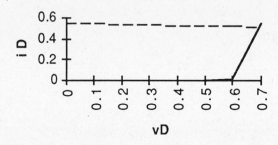

From the load line, we determine that

$$i_D \approx 0.51 \text{ A}$$
$$v_D \approx 0.7 \text{ V}$$

b) One possible BASIC program is the following:

```
10010  ISAT=1E-12
10020  KTOVERQ=.0259
10030  RT=22
10040  VT=12
10050  VD1=VT/2
10060  VD2=VT
10070  ID1=(VT-VD1)RT
10080  ID2=ISAT*(EXP(VD1KTOVERQ)-1)
10090  IF ABS(VD1-VD2)<.000001 THEN 10120
10100  IF ID1>ID2 THEN LET VD1=VD1+(VD2-VD1)/2: GOTO 10070
10110  IF ID2>ID1 THEN LET VD2=VD1: VD1=VD1/2: GOTO 10070
10120  PRINT "iD =";ID1,"vD =";VD1
10130  END
```

2.34

## Section 3: AC Network Analysis

**3.1** Find the rms voltage of the waveform of Figure P4.4:

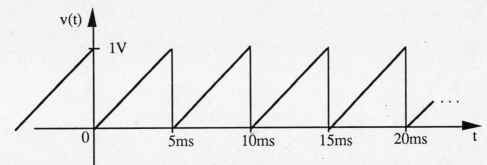

**Figure P4.4**

**Solution:** The voltage v(t) can be expressed as:
v(t) = 200t $\quad\quad$ 0 < t < 5 ms
The rms value of v(t) is

$$V_{rms} = \sqrt{\frac{1}{T} \int_0^T v^2(t)\, dt}$$

$$= \sqrt{\frac{1}{5\times10^{-3}} \int_0^{5\times10^{-3}} 200^2\, t^2\, dt}$$

$$= \sqrt{\frac{1}{5\times10^{-3}}\, 200^2\, \frac{(5\times10^{-3})^3}{3}}$$

$$= \frac{1}{\sqrt{3}} = 0.577 \text{ V}$$

**3.2**  Find the rms voltage of the waveform of Figure P4.5:

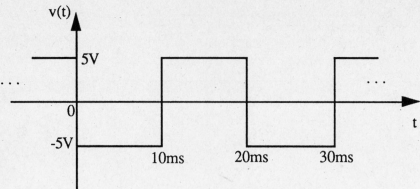

**Figure  P4.5**

**Solution:**

The rms voltage is:

$$v_{rms} = \sqrt{\frac{1}{0.02} \left[ \int_{0}^{0.01} (-5)^2 \, dt + \int_{0.01}^{0.02} (5)^2 \, dt \right]}$$

$$= \sqrt{50 \, (0.25 + 0.5 - 0.25)} = \sqrt{25} = 5 \text{ V}$$

3.2

**3.3**    a)  Find the rms voltage of the waveform of Figure P4.9:

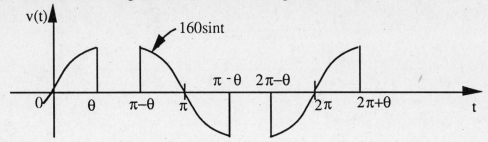

**Figure  P4.9**

b)  If $\theta = 30°$    find $V_{rms}$

c)  If $\theta = 60°$    find $V_{rms}$

d)  If $\theta = 90°$    find $V_{rms}$

**Solution:**

$$v_{rms} = \{ \frac{1}{2\pi} ( \int_0^\theta (160\sin t)^2 dt$$

$$+ \int_{\pi-\theta}^{\pi+\theta} (160\sin t)^2 dt + \int_{2\pi-\theta}^{2\pi} (160\sin t)^2 dt) \}^{1/2}$$

$$= \sqrt{\frac{1}{2\pi} (25600 ( 2\theta - \sin 2\theta))}$$

a) For $\theta = 30° = \frac{\pi}{6}$, we have

$$v_{rms} = \sqrt{\frac{1}{2\pi} (25600(\frac{2\pi}{6} - \sin (\frac{\pi}{3})))} \qquad = 27.2 \text{ V.}$$

b) For $\theta = 60° = \frac{\pi}{3}$,

$$v_{rms} = \sqrt{\frac{1}{2\pi} (25600(\frac{2\pi}{3} - \sin (\frac{2\pi}{3})))}$$

$$= 70.7 \text{ V}$$

c) For $\theta = 90° = \frac{\pi}{2}$

$$v_{rms} = \frac{160}{\sqrt{2}} = \sqrt{\frac{1}{2\pi}(25600(\pi - \sin 2\pi))}$$

$$= 113.1 \text{V}$$

**3.4** If a sine wave of current $i_C(t) = I_m \sin\omega t$ passes through a capacitor $C = 1F$, draw the voltage waveform.

**Solution:**

$$v_c(t) = \frac{1}{C} \int_{-\infty}^{t} i(\tau)\, d\tau$$

$$= \frac{I_m}{C} \int_{-\infty}^{t} \sin\omega\tau\, d\tau$$

$$= -\frac{I_m}{\omega} \cos\omega t$$

The voltage waveform is shown below.

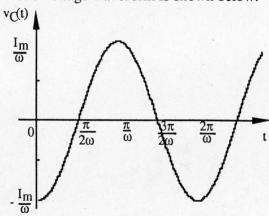

**3.5**    If a cosine wave of current $i_L(t) = I_m\cos\omega t$ flows through an inductor L = 1H, draw the voltage waveform.

**Solution:**

$$v = L\frac{di}{dt} = \frac{d}{dt}(I_m\cos\omega t)$$

$$= -I_m\,\omega\sin\omega t$$

The voltage waveform is shown below.

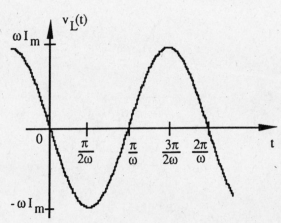

**3.6**  The voltage across a 20µF capacitor takes the values shown below.  Calculate the sinusoidal expression of current through the capacitor in each case.

    a)  $v_C(t) = 30\sin 50t$  V

    b)  $v_C(t) = 10\cos 30t$  V

    c)  $v_C(t) = -60\sin 10t$  V

    d)  $v_C(t) = 4\cos(100t + \frac{\pi}{2})$  V

**Solution:**

$$i_C = C \frac{dv_C}{dt}$$

a) $i_C(t) = 20 \times 10^{-6} \times \frac{d}{dt} (30 \sin 50t)$

$$= 20 \times 10^{-6} \times 30 \times 50 \times \cos 50t$$
$$= 30 \cos 50t \ mA$$

b) $i_C(t) = 20 \times 10^{-6} \times \frac{d}{dt} (10 \cos 30t)$

$$= 20 \times 10^{-6} \times 10 \times 30 \times (-\sin 30t)$$
$$= -6 \sin 30t \ mA$$

c) $i_C(t) = 20 \times 10^{-6} \times \frac{d}{dt} (-60 \sin 10t)$

$$= 20 \times 10^{-6} \times (-60) \times 10 \times \cos 10t$$
$$= -12 \cos 10t \ mA$$

d) Note that $\cos(x + \frac{\pi}{2}) = -\sin x$

$$\therefore \ i_C(t) = 20 \times 10^{-6} \times \frac{d}{dt} (-4 \sin 100t)$$

$$= 20 \times 10^{-6} \times (-4) \times 100 \times (\cos 100t)$$

$$= -8 \cos 100t \ mA$$

---

**3.7**   The current through a 10mH inductor takes the values shown below.  Calculate the sinusoidal expression of voltage across the inductor in each case:

a) $i_L(t) = 10\sin 50t \ mA$

b) $i_L(t) = 4\cos 200t \ mA$

c) $i_L(t) = 20\sin(10t + \frac{\pi}{4}) \ mA$

d) $i_L(t) = -6\sin 40t \ mA$

**Solution:**

$$v_L(t) = L \frac{di_L}{dt}$$

a) $v_L(t) = 10 \times 10^{-3} \frac{d}{dt} (10 \times 10^{-3} \sin 50t)$

$$= 100 \times 10^{-6} \times 50 \times \cos 50t$$
$$= 5 \cos 50t \ mV$$

b) $v_L(t) = 10 \times 10^{-3} \frac{d}{dt} (4 \times 10^{-3} \cos 200t)$

$$= 10 \times 10^{-3} \times 4 \times 10^{-3} \times 200 \times (-\sin 200t)$$
$$= -8 \sin 200t \ mV$$

c) $v_L(t) = 10 \times 10^{-3} \frac{d}{dt}(20 \times 10^{-3} \sin(10t + \frac{\pi}{4}))$

$$= 10 \times 10^{-3} \times 20 \times 10^{-3} \times 10 \times \cos(10t + \frac{\pi}{4})$$

$$= 2 \cos(10t + \frac{\pi}{4}) \ mV$$

d) $v_L(t) = 10 \times 10^{-3} \frac{d}{dt} (-6 \times 10^{-3} \sin 40t)$

$$= 10 \times 10^{-3} (-6 \times 10^{-3}) \times 40 \cos 40t$$
$$= -2.4 \times 10^{-3} \cos 40t$$
$$= -2.4 \cos 40t \ mV$$

**3.8** Calculate the energy stored in a 1000 μF capacitor at t=3ms by the waveform of Figure 4.17.

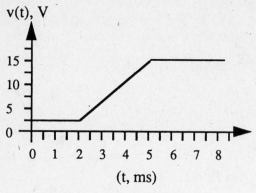

**Figure 4.17**

The voltage waveform of Figure 4.17 appears across a *50mH* inductor and a *1000μF* capacitor.

**Solution:**

The capacitor voltage in Drill Exercise 4.2 is

$$v_C(t) = 2.5 \text{ V} \qquad \text{for } 0 < t < 2 \text{ ms}$$
$$v_C(t) = 2.5 + 4167(t - 2 \text{ ms}) \quad \text{for } 2 < t < 6 \text{ ms}$$

Therefore, at $t = 3$ ms, the voltage is

$$v_C(3\text{ms}) = 2.5 + 4167 \times 1 \times 10^{-3}$$
$$= 6.67 \text{ V}$$

and the energy stored in the capacitor is:

$$W_C = \frac{1}{2}Cv_C^2$$
$$= \frac{1}{2} \times 0.001 \times (6.67)^2$$
$$= 22.24 \text{ mJ}$$

---

**3.9**    Calculate and plot the energy stored in the inductor for the case of Figure 4.16.

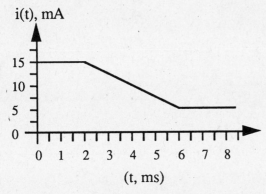

i(t), mA

(t, ms)

**Figure 4.16**

The current waveform shown in Figure 4.16 flows through a *50mH* inductor and a *1000μF* capacitor.

**Solution:**

$$\text{For the inductor, } w_L = \frac{1}{2} Li^2$$

$$\text{For } 0 < t < 2 \text{ ms,}$$
$$i(t) = 15 \text{ mA}$$

$$\therefore w_L = \frac{1}{2}(0.05)(0.015)^2 = 5.625 \ \mu J$$

$$\text{For } 2 \text{ ms} < t < 6 \text{ ms,}$$
$$i(t) = 0.015 - 2.5(t - 2 \text{ ms}) \text{ A}$$
$$= 0.02 - 2.5 \ t \text{ A}$$

$$\therefore w_L = \frac{1}{2}(0.05)(0.02 - 2.5 \ t)^2$$

$$= 0.025 \ (0.0004 - 0.1 \ t + 6.25 \ t^2) \text{ J}$$
$$\text{For } t > 6 \text{ ms,}$$
$$i(t) = 5 \text{ mA}$$

$$\therefore w_L = \frac{1}{2}(0.05)(0.005)^2 = 625 \text{ nJ}$$

---

**3.10** Calculate the energy stored in the capacitor shown in Figure P4.11.

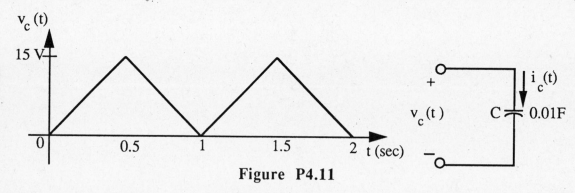

**Figure P4.11**

**Solution:**

$$W_C(t) = \frac{1}{2}Cv^2(t)$$

$$= \frac{1}{2} \times 0.01 \times (30t)^2 = 4.5t \quad 0 < t < 0.5 \text{ s}$$

$$W_C(t) = \frac{1}{2} \times 0.01 \times (30 - 30t)^2$$

$$= 4.5t^2 - 9t + 4.5 \qquad 0.5 < t < 1 \text{ s}$$

$$W_C(t) = \frac{1}{2} \times 0.01 \times (30t - 30)^2$$

$$= 4.5t^2 - 9t + 4.5 \qquad 1 < t < 1.5 \text{ s}$$

$$W_C(t) = \frac{1}{2} \times 0.01 \times (60 - 30t)^2$$

$$= 4.5t^2 - 18t + 18 \qquad 1.5 < t < 2$$

---

**3.11** Calculate the energy stored in inductor shown in Figure P4.13.

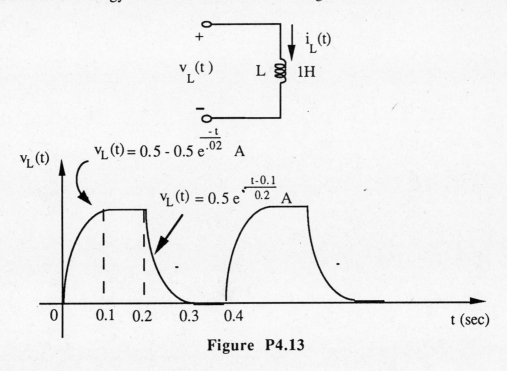

**Figure P4.13**

**Solution:**

For the inductor, $w_L = \frac{1}{2}Li_L^2$

For $0 < t < 0.1$ s

$$i_L(t) = 0.5t + 0.01e^{-t/0.02} - 0.01 \ \text{A}$$

$$\therefore \ w_L = \frac{1}{2}[0.5t + 0.01e^{-t/0.02} - 0.01]^2 \ \text{J}$$

For $0.1 < t < 0.2$ s

$$i_L = 0.49663t - 0.009596 \ \text{A}$$

$$\therefore \ w_L = \frac{1}{2}[0.49663t - 0.009596]^2 \ \text{J}$$

For $0.2 < t < 0.3$ s

$$i_L = 0.5t - 0.01e^{(t-0.3)/0.02} - 0.0102 \ \text{A}$$

$$\therefore \ w_L = \frac{1}{2}[0.5t - 0.01e^{(t-0.3)/0.02} - 0.0102]^2 \ \text{J}$$

For $0.3 < t < 0.4$ s

$$i_L = 0.1298 \ \text{A}$$

$$\therefore \ w_L = \frac{1}{2}[0.1298]^2$$

$$= 8.42 \ \text{mJ}$$

**3.12**  For the circuit shown in Figure P4.17, find $v_{out}$ by writing the differential equation. Assume that $v_{out}(t)$ has the form $v_{out} = A \sin\omega t + B \cos\omega t$ and using trigonometric identities find $v_{out}(t)$ in the form:

$$v_{out}(t) = D \cos(\omega t + \phi)$$

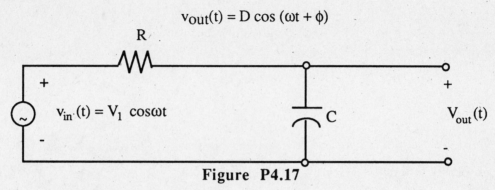

Figure  P4.17

**Solution:**

Applying KVL, we find that

$V_1 \cos\omega t = v_R + v_{out}$

where $v_R = Ri = R(C\frac{dv_{out}}{dt})$.  Thus,

$$RC\frac{dv_{out}}{dt} + v_{out} = V_1 \cos\omega t \qquad (1)$$

Assuming $v_{out}(t) = A \sin\omega t + B \cos\omega t$,
we have

$$\frac{dv_{out}}{dt} = A\omega \cos\omega t + (-B\omega) \sin\omega t$$

$$= A\omega \cos\omega t - B\omega \sin\omega t \qquad (2)$$

By replacing $\frac{dv_{out}}{dt}$ with (2) in (1), we have

$$RC (A\omega \cos\omega t - B\omega \sin\omega t) +$$

$$(A \sin\omega t + B \cos\omega t) = V_1 \cos\omega t$$

or

$$(RC\omega A + B) \cos\omega t +$$

$$(A - BRC\omega) \sin\omega t = V_1 \cos\omega t$$

Equating the coefficients of the sine and cosine terms, we have

1) $RC\omega A + B = V_1$

2) $A - BRC\omega = 0$,  that is  $A = BRC\omega$

Equation 1) now can be expressed as

1) $RC\omega (BRC\omega) + B = V_1$

$$B( (RC\omega)^2 + 1) = V_1$$

$$B = \frac{V_1}{(RC\omega)^2 + 1}$$

3.16

$$A = \frac{V_1 \, RC\omega}{(RC\omega)^2 + 1}$$

and therefore

$$v_{out}(t) = \frac{V_1 RC\omega}{(RC\omega)^2 + 1} \sin\omega t$$

$$+ \frac{V_1}{(RC\omega)^2 + 1} \cos\omega t$$

Applying the identity:

$$K_1 \cos\omega t + K_2 \sin\omega t = K\cos(\omega t - \phi), \text{ with}$$

$$K = \sqrt{K_1{}^2 + K_2{}^2}$$

and

$$\phi = \tan^{-1}\!\left(\frac{K_2}{K_1}\right)$$

we determine the coefficient

$$K = \sqrt{A^2 + B^2}$$

$$= \sqrt{(\frac{V_1{}^2}{(\omega RC)^2 + 1})^2 \, [(\omega RC)^2 + 1]}$$

$$= \frac{V_1}{\sqrt{(\omega RC)^2 + 1}}$$

$$\phi = \tan^{-1}\!\left(\frac{A}{B}\right) = \tan^{-1}(\omega RC)$$

Therefore, the complete expression for the voltage is

$$v_{out}(t) = \frac{V_1}{\sqrt{(\omega RC)^2 + 1}} \cos[\omega t - \tan^{-1}\omega RC)]$$

---

**3.13** Write and solve the differential equation for $v_{out}(t)$ for the circuit shown in Figure P4.17 if $R = 10\Omega$, $C = 20\mu F$, and $v_{in}(t) = 10\cos(2500t)$.

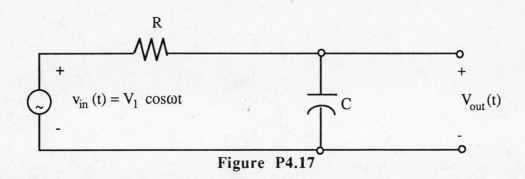

**Figure P4.17**

**Solution:**
By KVL we have:

$$10\cos2500t = 10\times20\times10^{-6}\frac{dv_{out}}{dt} + v_{out}$$

In standard form:

$$\frac{dv_{out}}{dt} + 5000v_{out} = 50000\cos2500t$$

Assume $v_{out} = A\sin2500t + B\cos2500t$

$$\frac{dv_{out}}{dt} = 2500A\cos2500t - 2500B\sin2500t$$

Substituting,

$$2500A\cos2500t - 2500B\sin2500t +$$
$$+ 5000A\sin2500t + 5000B\cos2500t$$
$$= 50000\cos2500t$$

Grouping terms:

$$(2500A + 5000B)\cos2500t +$$
$$+ (-2500B + 5000A)\sin2500t$$
$$= 50000\cos2500t$$

which leads to:

$$2500A + 5000B = 50000$$
and $\qquad 5000A - 2500B = 0$
or $\qquad\qquad B = 2A$
Therefore,
$$2500A + 5000(2A) = 50000$$
$$A = 4 \quad B = 8$$

and $\quad v_{out} = 4\sin2500t + 8\cos2500t$

---

**3.14** For the circuit shown in Figure P4.23, find the expression of $v_C(t)$ by writing the differential equation.

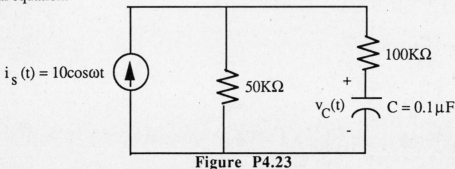

**Figure P4.23**

**Solution:**

By KCL: $10\cos\omega t = i_R + i_C$

$$i_C = C\frac{dv_C}{dt} = 0.1\times10^{-6}\frac{dv_C}{dt}$$

$$i_R = \frac{v_R}{50000}$$

$$v_R = 100\times10^3\, i_C + v_C$$

$$= 0.01\frac{dv_C}{dt} + v_C$$

Substituting into the first equation:

$$5\times10^5\cos\omega t = 0.01\frac{dv_C}{dt} + v_C + 0.5\times10^{-2}\frac{dv_C}{dt}$$

$$5\times10^5\cos\omega t = 0.015\frac{dv_C}{dt} + v_C$$

In standard form:

$$\frac{dv_C}{dt} + 66.67v_C = 33.3\times10^6\cos\omega t$$

---

**3.15** For the circuit shown in Figure P4.17, find $v_{out}$ using phasor techniques. Find $v_{out}(t)$ in the form:

$$v_{out}(t) = D\cos(\omega t + \phi)$$

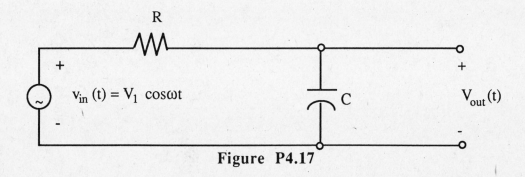

$$v_{in}(t) = V_1\cos\omega t$$

**Figure P4.17**

**Solution:**
By the voltage divider, we have

$$\mathbf{V}_{out}(\omega) = \frac{\dfrac{1}{j\omega C}}{R + \dfrac{1}{j\omega C}} \mathbf{V}_{in}(\omega)$$

$$= \frac{1}{1 + j\omega C R} V_1 \angle 0°$$

$$= \frac{V_1}{\sqrt{1 + (\omega CR)^2}} \angle -\tan^{-1}\omega CR$$

Thus, the output voltage expressed in time domain form is:

$$v_{out}(t) = \frac{V_1}{\sqrt{1 + (\omega CR)^2}} \cos(\omega t - \tan^{-1}\omega CR)$$

---

**3.16** Write the differential equation for the circuit shown in Figure P4.18 using phasor techniques. Assume $\omega = 377$ rad/s.

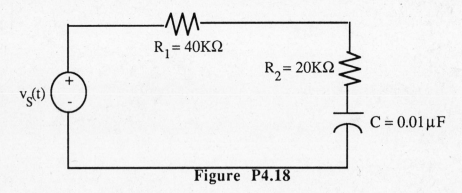

**Figure P4.18**

**Solution:**

In the circuit of Figure 4.18, finding the series current is sufficient to solve for any of the voltages in the series circuit; thus, we seek the current:

$$\mathbf{I}(\omega) = \frac{\mathbf{V}_S(\omega)}{R_1 + R_2 + \dfrac{1}{j\omega C}}$$

For an arbitrary voltage

$$\mathbf{V}_S(\omega) = V \angle \phi$$

3.21

the current may be found as follows:

$$\mathbf{I}(\omega) = V\angle\phi \times \frac{1}{R_1 + R_2 + 1/j\omega C}$$

$$= V\angle\phi \times \frac{j\omega C}{(R_1+R_2)j\omega C+1}$$

$$= \frac{V\omega C}{\sqrt{[(R_1+R_2)\omega C]^2+1}} \angle\{\phi+90°-\tan^{-1}[(R_1+R_2)\omega C]\}$$

For example, if $v_S(t) = 5\cos(377t)$, we have

$$\mathbf{V}_C(\omega) = \frac{1}{j\omega C} \times \mathbf{I}(\omega) = 4.88\angle{-12.75°}$$

and

$$v_C(t) = 4.88\cos(377t - 12.75°)$$

---

**3.17** Using phasor techniques, find the output voltage, $v_{out}(t)$, for the circuit shown in Figure P4.19 by assuming the solution has the form

$$v_{out}(t) = D\cos(\omega t + \phi)$$

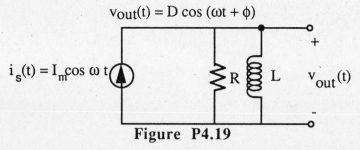

**Figure P4.19**

**Solution:**
The impedance of the circuit is:

$$Z(\omega) = \frac{R\ Z_L}{R + Z_L} = \frac{R\ j\omega L}{R + j\omega L}$$

The output voltage is:

$$\mathbf{V_{out}}(\omega) = \mathbf{I_S}(\omega) \times Z(\omega) = \frac{R j\omega L}{R + j\omega L}\,\mathbf{I_S}(\omega)$$

Thus, if $\mathbf{I_S}(\omega) = I_m \angle 0$,
then

$$\mathbf{V_{out}}(\omega) = \frac{R\omega L I_m}{\sqrt{R^2 + \omega^2 L^2}} \angle\,[90° - \tan^{-1}(\omega L/R)]$$

$$v_{out}(t) = \frac{R\omega L I_m}{\sqrt{R^2 + \omega^2 L^2}} \cos(\omega t + 90° - \tan^{-1}(\omega L/R))$$

---

**3.18** a) Find the current flowing in the series RLC circuit shown in Figure P4.21 using phasor techniques and assuming the solution has the form:

$$i(t) = (A\ \sin\omega t + B\ \cos\omega t)\ A$$

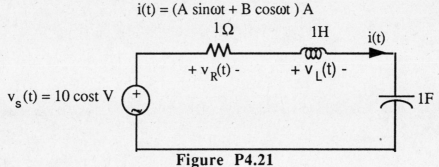

**Figure P4.21**

b) What is the inductor voltage $v_L(t)$?
c) What is the resistor voltage $v_R(t)$?

**Solution:**

a)

$$I(\omega) = \cfrac{1}{R + j\omega L + \cfrac{1}{j\omega C}} \times 10\angle 0 = \frac{j\omega C}{j\omega CR + -\omega^2 LC + 1} \times 10\angle 0$$

$$= \frac{10\omega C}{\sqrt{(\omega CR)^2 + (1-\omega^2 LC)^2}} \angle 90° - \tan^{-1}[(\omega CR)/(1 -\omega^2 LC)]$$

or

$$i(t) = \frac{10\omega C}{\sqrt{(\omega CR)^2 + (1-\omega^2 LC)^2}}$$
$$\times \cos(\omega t + 90° - \tan^{-1}[(\omega CR)/(1 -\omega^2 LC)])$$
$$= 10 \cos t \ A$$

b)  We can obtain the inductor voltage from the above expressions by observing that
$$V_L(\omega) = j\omega L \ I(\omega) = \omega\angle 90° \ I(\omega)$$
Therefore,
$$v_L(t) = 10 \cos(t + 90°) = -10 \sin t \ V$$

c)  Similarly,   $V_R(\omega) = R \ I(\omega) = I(\omega)$
$$\text{So,} \ \ v_R(t) = 10 \cos t \ V$$

---

**3.19**  Find the sinusoidal steady-state outputs for each of the circuits shown in Figure P4.29.

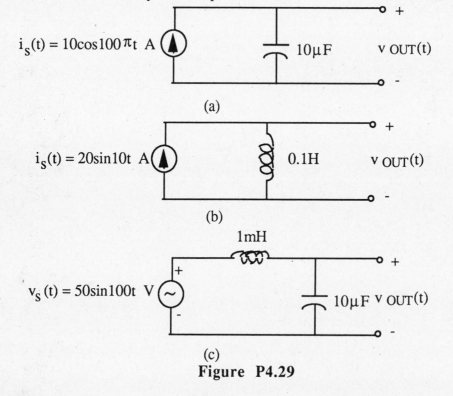

(a)

(b)

(c)

**Figure  P4.29**

**Solution:**
For circuit a):

$$\mathbf{V}_{out} = 10\angle 0° \times (\frac{1}{\pi 10^{-3}}\angle\text{-}90°)$$

$$= \frac{10000}{\pi}\angle\text{-}90° = 3183\angle\text{-}90°$$

$$v_{out}(t) = 3183\cos(100\pi t - 90°) \text{ V}$$

For circuit b):

$$\mathbf{V}out = 20\angle 0° \times 1\angle 90° = 20\angle 90°$$
$$v_{out}(t) = 20\sin(10t + 90°) = 20\cos 10t \quad V$$

For circuit c):

$$\mathbf{V}_{out} = \frac{-j1000}{j0.1-j1000} \times 50\angle 0° \approx 50\angle 0°$$
$$v_{out}(t) = 50\sin 100t \quad V$$

---

**3.20** Find the voltage drop $v_{CL}$ shown in Figure P4.32.

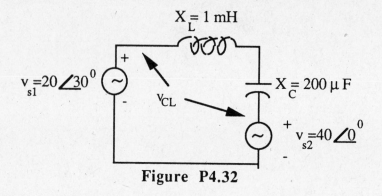

**Figure P4.32**

**Solution:**
From KVL we have

$$40\angle 0° = \mathbf{V}_{CL} + 20\angle 30°$$
$$\mathbf{V}_{CL} = 40\angle 0° - 20\angle 30°$$
$$= (40 - 17.32) + j\,(0-10)$$
$$= 22.679 - j\,10 = 24.8\angle -23.8°$$
$$v_{CL}(t) = 24.8\cos(\omega t - 23.8°) \text{ V}$$

---

**3.21** Compute the Thevenin impedance seen by resistor $R_2$ in Figure P4.37.

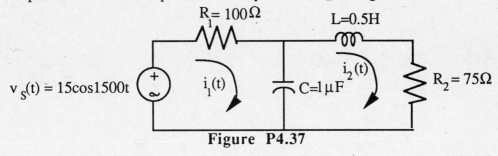

$v_S(t) = 15\cos 1500t$, $R_1 = 100\,\Omega$, $L = 0.5\text{H}$, $C = 1\,\mu\text{F}$, $R_2 = 75\,\Omega$, $i_1(t)$, $i_2(t)$

**Figure P4.37**

**Solution:**

$$Z_T = 100 \parallel (-\frac{j2000}{3}) + j750$$
$$= 97.7995 - j14.6699 + j750$$
$$= 97.8 + j735.33 = 741.8\angle 82.4°\ \Omega$$

---

**3.22** Compute the Thevenin Voltage seen by resistor $R_2$ if $v_S(t) = 2\cos(2t)$, $R_1 = 4\Omega$, $R_2 = 4\Omega$, $L=2H$ and $C=\frac{1}{4}$ F. The circuit is shown in Figure P4.36.

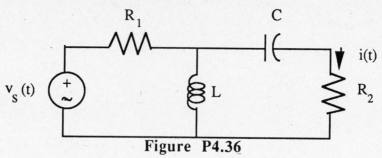

**Figure P4.36**

**Solution:**

From the voltage divider, we have

$$V_T = \frac{j4}{4 + j4} 2\angle 0° = (1 + j)$$

$$= \sqrt{2}\angle 45° = 1.414\angle 45° \text{ V}$$

$$v_T(t) = 1.414\cos(2t + 45°) \text{ V}$$

**3.23** Find the Norton equivalent circuit seen by resistor $R_2$ if $v_S(t) = 2\cos(2t)$, $R_1 = 4\Omega$, $R_2 = 4\Omega$, $L = 2H$ and $C = \frac{1}{4}$ F. The circuit is shown in Figure P4.36.

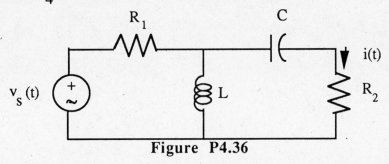

**Figure P4.36**

**Solution:**
From the result of Problem 4.64, we have $Z_T = 2\ \Omega$. From the current divider, we have

$$I_N = \frac{j4}{j4 - j2} I = 2I$$

and

$$j4 \parallel -j2 = \frac{(-j2)(j4)}{j2} = -j4$$

The current is

$$I = \frac{2\angle 0°}{4 - j4} = \frac{\sqrt{2}}{4}\angle 45° = 0.35355\angle 45°\ A$$

Therefore,

$$I_N = 2I = 2(\frac{\sqrt{2}}{4}\angle 45°) = 0.707\angle 45°\ A$$

# Section 4: AC Power

**4.1** The waveform of the voltage source and the circuit is shown in Figure P5.4.

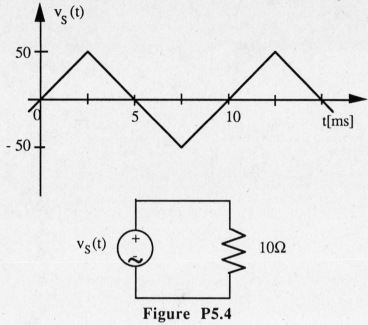

**Figure P5.4**

a) Find the rms voltage, $V_S$, of $v_S(t)$.

b) If the voltage, $v_S(t)$ is applied to a 10 $\Omega$ resistor, what is the average power supplied to the resistor?

**Solution:**

(a) $V_S^2 = \frac{1}{T} \int_0^T v_s^2(t)\, dt$

$= 2 \times \frac{1}{T} \{ \int_0^{0.0025} 4 \times 10^8 x^2 dt +$

$+ \int_{2.5 \times 10^{-3}}^{5 \times 10^{-3}} (4 \times 10^8 x^2 - 4 \times 10^6 x + 10^4) dt \}$

$= 2 \times 417 = 834$

Therefore, the rms value is

$$V_S = 28.87 \text{ V}$$

(b) $P = \frac{V_S^2}{R} = \frac{834}{10}$

$= 83.4 \text{ W}$

---

**4.2**  Find the instantaneous real and reactive power supplied by the source in the circuit shown in Figure P5.7.  $\omega = 337$ rad/sec.

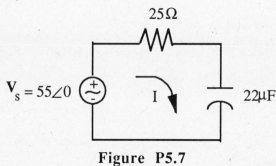

**Figure  P5.7**

**Solution:**

$V_s = 55\angle 0°$ V

$$Z_C = \frac{1}{j\omega C} = -j120.57\ \Omega$$

$$I = \frac{V_s}{R + Z_C} = \frac{55}{25 - j120.57}$$

$$= 0.4467\angle 78.29°\ A$$

or:

$$i(t) = 0.4467\sqrt{2}\cos(377t + 78.29°)\ A$$

The instantaneous real power is:

$p_R = v_s i = i^2 R$

$$= (0.4467\sqrt{2})^2 \cos^2(377t + 78.29°)\times 25$$

$$= 4.99 + 4.99\cos(754t + 156.58°)\ W$$

The voltage across the capacitor is:

$$v_c = \frac{1}{C}\int i\ dt + v_c(0) \qquad v_c(0) = 0\ V$$

$$v_c = 76.16\sin(377t + 78.29°)\ V$$

The instantaneous reactive power is:

$p_X = iv_c = 0.4467\sqrt{2}\cos(377t + 78.29°)$

$\qquad \times 76.16\sin(377t + 78.29°)$

$= 48.11\ \sin(754t + 156.58°)\quad VAR$

---

**4.3** Use complex power to determine the parallel circuit equivalent of a series R-L circuit, that is, find $R_2$ and $L_2$ in terms of $R_1$, $L_1$ and $\omega$ by computing the complex power for each circuit (for arbitrary voltage or current), and equating the two. The circuits are shown in Figure P5.9.

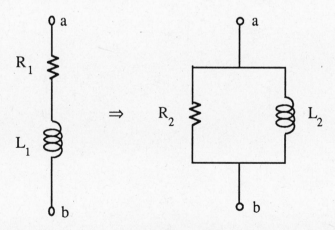

**Figure P5.9**

**Solution:**

The equivalent impedance $Z_2$ is

$$Z_2 = \frac{R_2 \times j\omega L_2}{R_2 + j\omega L_2}$$

$$= j\frac{R_2^2\omega L_2 + R_2\omega^2 L_2^2}{R_2^2 + \omega^2 L_2^2}$$

The equivalent impedance $Z_1$ is

$$Z_1 = R_1 + j\omega L_1$$

To find the equivalent components (in the sense of complex power usage), we set

$$\frac{R_2\omega^2 L_2^2}{R_2^2 + \omega^2 L_2^2} = R_1$$

$$\frac{R_2^2\omega L_2}{R_2^2 + \omega^2 L_2^2} = \omega L_1$$

Solving the above equations, we have

$$L_2 = \frac{R_1^2 + \omega^2 L_1^2}{\omega^2 L_1}$$

$$R_2 = \frac{R_1^2 + \omega^2 L_1^2}{R_1}$$

**4.4** A circuit is shown in Figure P5.13 to contain 3 elements in parallel. Two tests are performed on the circuit, the first at 1000 Hz and the second at 100 Hz. The following data are recorded from the two tests:

1,000-Hz:  $v_{in}(t) = 100 \sin(6{,}283t)$ V

$i_{in}(t) = 44.51 \sin(6{,}283t + 77°)$ A

$P_{in} = 500$ W

100-Hz:  $v_{in}(t) = 50 \sin(628t)$ V

$i_{in}(t) = 75.6 \sin(628t - 85.3°)$ A

$P_{in} = 155$ W

Determine the value of the circuit elements.

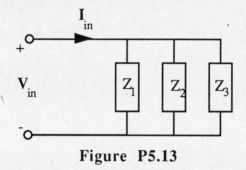

**Figure  P5.13**

**Solution:**

From the circuit, we have

$$I_{in} = \frac{V_{in}}{R} + \frac{V_{in}}{j6283L} + \frac{V_{in}}{1/j6283C} \qquad (1)$$

$$I'_{in} = \frac{V'_{in}}{R} + \frac{V'_{in}}{j628L} + \frac{V'_{in}}{1/j628C} \qquad (2)$$

From $P_{in} = \frac{(V_{in}/\sqrt{2})^2}{R} = 500$, we have

$$R = 10 \ \Omega$$

Equations (1), (2) become

$$\frac{44.51}{\sqrt{2}}\angle 77° = 7.1 + j30.7$$

$$= \frac{100}{\sqrt{2}10} + \frac{100}{\sqrt{2}j6283L} + \frac{100j6283C}{\sqrt{2}}$$

$$\frac{75.6}{\sqrt{2}}\angle -85.3° = 4.38 - j53.28$$

$$= \frac{50}{\sqrt{2}10} + \frac{50}{\sqrt{2}j628L} + \frac{50j628C}{\sqrt{2}}$$

Solving the above equations (imaginary parts), we have L = 1.07 mH and C = 93.1 μF

---

**4.5**    A resistance and a capacitor are connected in series across a 220V rms 60Hz source. The resistance is adjusted so that an ammeter in the circuit indicates 8A rms and a wattmeter indicates 250W.  Find the resistance and the capacitance of the circuit.

**Solution:**

The power indicated by the wattmeter is:
$$P = VI\cos\theta = 220 \times 8 \times \cos\theta = 250 \ W$$

Thus, the power factor is
$$\cos\theta = 0.142$$

From $P = I^2 R\cos\theta = 250$, we have
$$R = \frac{250}{64 \times 0.142} = 27.5 \ \Omega$$

The angle of the load impedance is such that
$$\frac{1/377C}{27.5} = \tan\ (\cos^{-1}0.142) = 6.97$$

Therefore, the value of the capacitor is
$$C = \frac{1}{377 \times 27.5 \times 6.97} = 13.8 \ \mu F$$

---

**4.6** If we want the circuits shown in Figure P5.16 to be at unity power factor, find $C_P$ and $C_S$.

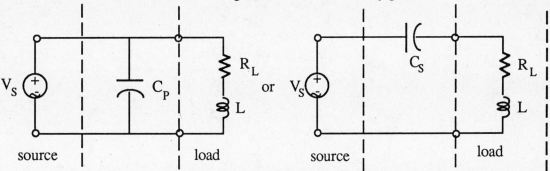

$$R_L = 5\Omega, \quad X_L = 5\Omega, \quad v_S(t) = 100\sin(377t)$$

**Figure P5.16**

**Solution:**

(a) The load current in the parallel circuit is

$$I_L = \frac{100/\sqrt{2}}{5 + j5} = 10\angle\text{-}45° \text{ A}$$

The reactive power in the inductor is

$$Q_L = I_L^2 X_L = 10^2 \times 5 = 500 \text{ VAR}$$

Thus, the capacitive reactance required to cancel the reactive power in the inductor is

$$X_C = \frac{V_S^2}{Q_L} = 10 \text{ }\Omega$$

The required capacitor is

$$C = \frac{1}{377 X_C} = 265.3 \text{ }\mu\text{F}$$

(b) In the series circuit, we can cancel the inductive reactance by setting

$$j\omega L + \frac{1}{j\omega C} = 0,$$

resulting in

$$C = \frac{1}{\omega^2 L} = \frac{1}{\omega X_L} = \frac{1}{377 \times 5} = 530.5 \text{ }\mu\text{F}$$

4.8

**4.7** The transformer shown in Figure P5.19 has several sets of windings on the secondary side. The windings have turns ratios:

a: 15:1

b: 4:1

c: 12:1

d: 18:1

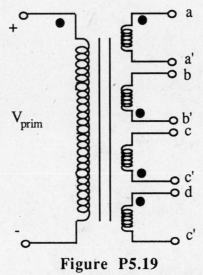

**Figure P5.19**

If $V_{prim}$ = 120V find and draw the connections which will allow you to construct the following voltage sources.

a) 24.67V∠0°

b) 35.67V∠0°

c) 18V∠0°

d) 54.67V∠180°

**Solution:**

The transformer direct output voltages are:

$$V_{aa'} = \frac{120}{15} = 8 \ V; \ V_{bb'} = \frac{120}{4} = 30 \ V$$

$$V_{cc'} = \frac{120}{12} = 10 \ V; \ V_{dd'} = \frac{120}{18} = 6.67 \ V$$

The connections required to obtain the desired voltages are shown below.

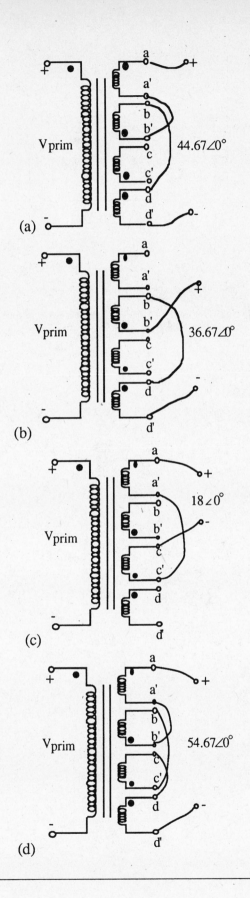

**4.8** The circuit shown in Figure P5.24 shows the use of ideal transformers for impedance matching.

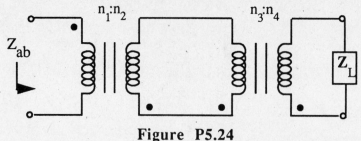

**Figure P5.24**

The problem is that you have a limited choice of turns ratios for the transformers available. The turns ratios are: 2:1, 7:2, 120:1, 3:2, and 6:1. If $Z_L$ is $475\Omega\angle-25°$ and $Z_{ab}$ must be $267\angle-25°$, find the combination of transformers which will provide this impedance. (You may assume polarity markings are easily changed on these transformers.)

**Solution:**

From $\frac{475}{267} = 1.779$, we have

$$\left(\frac{n_4}{n_3}\right)^2\left(\frac{n_2}{n_1}\right)^2 = 1.779$$

That is

$$\left(\frac{n_4}{n_3}\right)\left(\frac{n_2}{n_1}\right) = 1.33379 \approx 1.334$$

From 2:1 and 2:3, we have

$$\frac{2}{1}\times\frac{2}{3} = 2\times0.667 = 1.334$$

Therefore, we can choose the transformers with the turns ratio of 2:1 and 2:3 to match this impedance.

**4.9** In order to correct the power factor problems of the motor in the previous problem, the company has decided to install capacitors as shown in Figure P5.34:

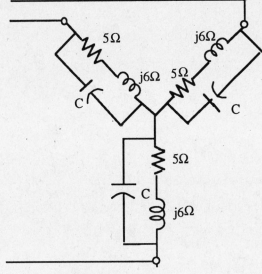

**Figure P5.34**

a) What capacitance must be installed to achieve unity power factor if the line frequency is 60 Hz?

b) Repeat part a) if we wish the power factor to be 0.85.

**Solution:**

This problem can be solved on a per-phase basis, due to its symmetry.

(a) The reactive power per phase is

$$Q = 120 \times 15.36 \times \sin(50.19°)$$

$$= 1415.9 \text{ VAR}$$

To achieve a unit power factor, we need

$$X_C = \frac{V^2}{Q} = 10.17 \ \Omega$$

The capacitance therefore is

$$C = \frac{1}{\omega X_C} = 260.8 \ \mu F$$

(b) If pf is 0.85, the impedance angle is

$$\theta = \cos^{-1}(0.85) = 31.79°$$

From the power triangle, the reactive power is

$$Q = P \times \tan(31.79°)$$

$$= 120 \times 15.36 \times \cos(50.19°) \times 0.62$$

$$= 731.46 \text{ VAR}$$

Therefore, we can write

$$X_C = \frac{V^2}{Q} = 19.69 \ \Omega$$

The value of the capacitor is

$$C = \frac{1}{\omega X_C} = 134.7 \ \mu F$$

**4.10**  The voltage supplied to an industrial plant can be viewed as a single 3 phase supply with line voltage 208V. This supply does not supply power to just one piece of machinery. It must supply power to several different motors which are not all identical, nor are they connected in the same manner, and are switched on and off. The plant we are considering has 3 types of 3 phase motors.

A Δ connected motor (load) with the impedances shown in Figure P5.36(a) when running under normal conditions.

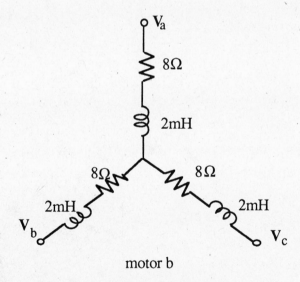

motor a

**Figure  P5.36  (a)**

A wye connected load with the following impedances shwon in Figure P5.36(b) when running under normal conditions.

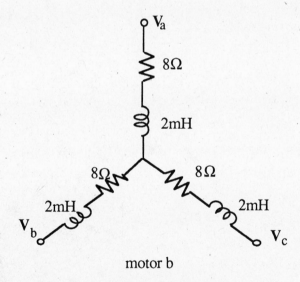

motor b

**Figure  P5.36  (b)**

A second wye connected load with the impedances shown in Figure P5.36(c) when running under normal conditions.

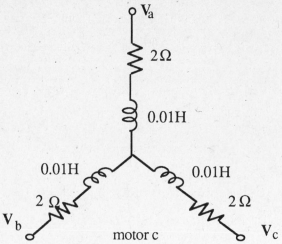

**Figure  P5.36  (c)**

The plant contains:    8 of motor  type a

15 of motor type b

and 30 of motor type c.

During a typical shift of 8 hours the plant runs:

6 motor  types a    8 hrs

2 motor  types a    4 hrs

1 motor  type  b    8 hrs

14 motor  types b    2 hrs

30 motor  types  b  6 hrs

a)  Find the average power delivered to each of the 3 motor types.

b)  Find the reactive power delivered to each of the 3 motor types.

c)  Which of the motors has the best (closest to unity) power factor?

d)  From the data given for atypical shift and the motors running under normal conditions, find the energy in kilowatt-hours consumed by the plant.

**Solution:**

(a)  Motor "a" can be modeled as a wye load

$$Z_Y = \frac{Z_\Delta}{3} = \frac{5 + j377 \times 8 \times 10^{-3}}{3} = 1.67 + j1 \ \Omega$$

The current is

$$I_a = \frac{208/\sqrt{3}}{1.67 + j1} = 61.69 \angle\text{-}30.9° \ A$$

The real power dissipated by a single motor of type "a" is:

$$P_a = \sqrt{3} \times 208 \times 61.69 \cos(30.9°) = 19{,}070 \ W$$

The energy consumed by motor-a is

$$6 \times P_a \times 8 + 2 \times P_a \times 4 = 1{,}067{,}900 \ \text{W-hr}$$

Therefore, the average power delivered to motor-a in a shift (8 hours) is

$$P_a' = \frac{1067900}{8} = 133.49 \ kW$$

The current in motor "b" is

$$I_b = \frac{208/\sqrt{3}}{8 + j377 \times 2 \times 10^{-3}}$$
$$= 14.945 \angle\text{-}5.38° \ A$$

The real power dissipated by a single motor of type "b" is:

$$P_b = \sqrt{3} \times 208 \times 14.945 \cos(5.38°) = 5.36 \ kW$$

The energy consumed by motor "b" is

$$1 \times P_b \times 8 + 14 \times P_b \times 2 = 192980 \ \text{W-hr}$$

Therefore, the average power delivered to motor "b" in a shift is:

$$P_b' = \frac{192980}{8} = 24.122 \ kW$$

The current in motor "c" is:

$$I_c = \frac{208/\sqrt{3}}{2 + j377 \times 0.01} = 28.14 \angle\text{-}62.1° \ A$$

The real power dissipated by a single motor of type "c" is:

$$P_c = \sqrt{3} \times 208 \times 28.14 \cos(62.1°) = 4.75 \ kW$$

The energy consumed by motor-c is

$$30 \times P_c \times 6 = 855{,}190 \ \text{W-hr}$$

Therefore, the average power delivered to motor "c" in a shift is

$$P_c' = \frac{855190}{8} = 106.9 \ kW$$

(b)  From the power triangle, we have

$$\frac{Q}{P} = \tan(\theta)$$

For motor "a", we have $P_a' = 133.49$ kW and $\theta = 30.9°$; therefore, the reactive power is:

$$Q_a = P_a' \times tg30.9° = 79.89 \text{ kVAR}$$

For motor "b", we have $P_b' = 24.122$ kW and $\theta = 5.38°$; therefore, the reactive power is:

$$Q_b = P_b' \times tg5.38° = 2.27 \text{ kVAR}$$

For motor "c", we have $P_c' = 106.9$ kW and $\theta = 62.1°$,

Therefore, the reactive power is

$$Q_c = P_c' \times \tan(62.1°) = 201.9 \text{ kVAR}$$

(c) Motor "b" has the best power factor

$$pf_b = \cos 5.38° = 0.996$$

(d) The total real power is

$$P = P_a' + P_b' + P_c'$$
$$= 133.49k + 24.122k + 106.9k = 264.5 \text{ kW}$$

The energy in kilowatt-hours consumed by the plant for a shift is

$$8 \times P = 2116.1 \text{ kW-hr}$$

# Section 5: Frequency Response and Transient Analysis

**5.1**    a)   Determine the frequency response $\dfrac{V_O(j\omega)}{V_{in}(j\omega)}$ for the circuit of Figure P6.1 for

     $C = 10\mu F$    and $R = 10k\Omega$.

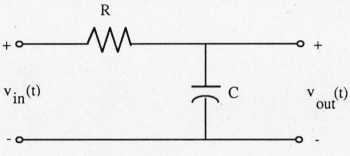

**Figure P6.1**

b) Plot the magnitude and phase of the circuit for frequencies between 1 and 100 $\dfrac{rad}{sec}$ on

     graph paper with a linear scale for frequency.

c) Repeat part b using semilog paper. (Place frequency on the log axis).

d) Plot magnitude response on semilog paper with magnitude in dB .

**Solution:**

(a) $\dfrac{V_{out}}{V_{in}} = \dfrac{1/j\omega C}{R + 1/j\omega C} = \dfrac{1}{j\omega RC + 1}$

$$\left|\dfrac{V_{out}}{V_{in}}\right| = \dfrac{1}{\sqrt{1 + 0.01\omega^2}}$$

$$\phi(\omega) = -\arctan(0.1\omega)$$

(b)

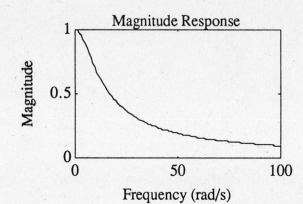

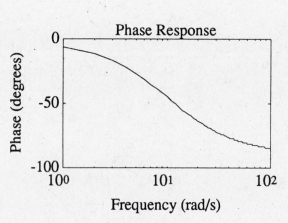

(c)

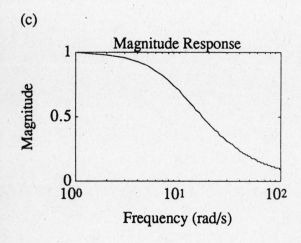

(d)

Magnitude Response

Magnitude (dB)

0

-50

$10^0$          $10^1$          $10^2$

Frequency (rad/s)

---

**5.2**   a)  Determine the frequency response $\dfrac{V_O(j\omega)}{V_{in}(j\omega)}$ for the circuit of Figure P6.4 for

$C = 100\mu F$    and $R = 1,000\Omega$.

b)  Plot the magnitude and phase of the circuit for frequencies between

1 and 100 $\dfrac{rad}{sec}$ on graph paper with a linear scale for frequency.

c)  Repeat part b using semilog paper.  (Place frequency on the log axis).

d)  Plot magnitude response on semilog paper with magnitude in dB .

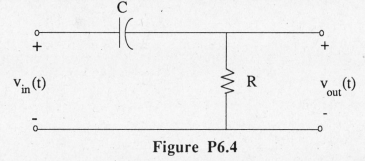

**Figure P6.4**

5.3

## Solution:

(a)   $\dfrac{V_{out}}{V_{in}} = \dfrac{j\omega/10}{1 + j\omega/10}$

$$\left|\frac{V_{out}}{V_{in}}\right| = \frac{0.1\omega}{\sqrt{1 + 0.01\omega^2}}$$

$$\phi(\omega) = 90° - \arctan(0.1\omega)$$

(b)

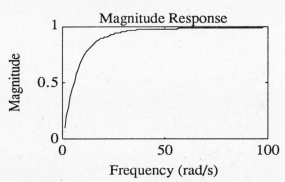

(c)

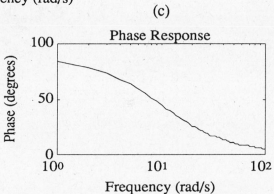

(d)

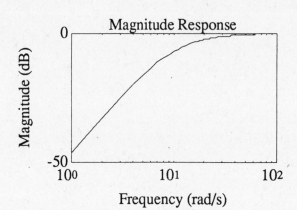

**5.3**  a)  Determine the frequency response $\dfrac{V_0(j\omega)}{V_{in}(j\omega)}$ for the circuit of Figure P6.4 for

$C = 10\mu F$  and $R = 10K\Omega$.

b)  Plot the magnitude and phase of the circuit for frequencies between
1 and $100 \dfrac{rad}{sec}$ on graph paper with a linear scale for frequency.

c)  Repeat part b using semilog paper.  (Place frequency on the log axis).

d)  Plot magnitude response on semilog paper with magnitude in dB .

Repeat Problem 6.5 for $C = 10\mu F$ and $R = 10k\Omega..$

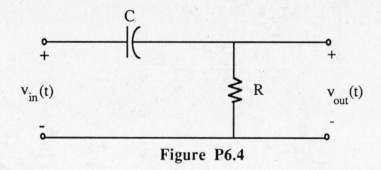

**Figure  P6.4**

**Solution:**

(a)
$$\frac{V_{out}(j\omega)}{V_{in}(j\omega)} = \frac{j\omega/10}{1 + j\omega/10}$$

$$\left|\frac{V_{out}}{V_{in}}\right| = \frac{0.1\omega}{\sqrt{1 + 0.01\omega^2}}$$

$$\phi(\omega) = 90° - \arctan(0.1\omega)$$

(b)

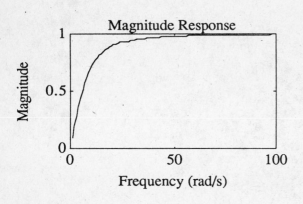

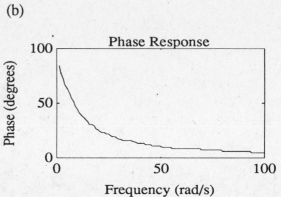

(c)

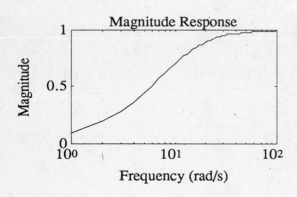

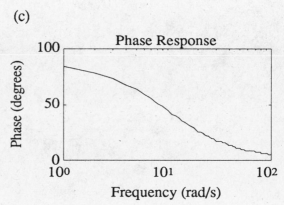

(d)

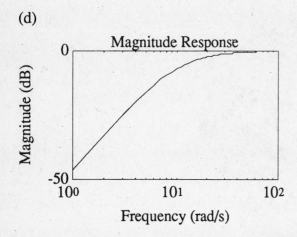

**5.4** a) Determine the frequency response $\dfrac{V_O(j\omega)}{V_{in}(j\omega)}$ for the circuit of Figure P6.7.

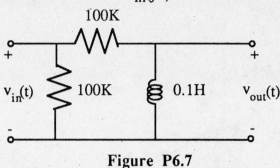

**Figure P6.7**

b) Plot the magnitude and phase of the circuit for frequencies between $100k \dfrac{rad}{sec}$ and $10M \dfrac{rad}{sec}$ on graph paper with a linear scale for frequency.

c) Repeat part b using semilog paper. (Place frequency on the log axis).

d) Plot magnitude response on semilog paper with magnitude in dB .

**Solution:**

(a) First, we find the Thèvenin equivalent circuit seen by the inductor,

$$R_T = 100 \text{ k} \parallel 100 \text{ k} = 50 \text{ k}\Omega$$

$$v_{oc} = \frac{100 \text{ k}}{100 \text{ k} + 100 \text{ k}} v_{in} = \frac{v_{in}}{2}$$

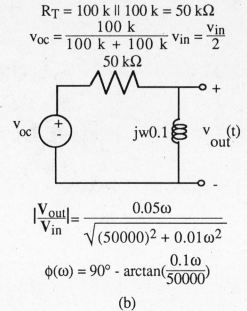

$$\left|\frac{V_{out}}{V_{in}}\right| = \frac{0.05\omega}{\sqrt{(50000)^2 + 0.01\omega^2}}$$

$$\phi(\omega) = 90° - \arctan\left(\frac{0.1\omega}{50000}\right)$$

(b)

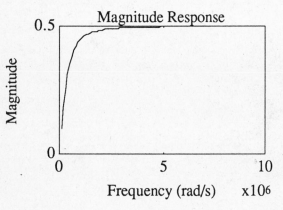

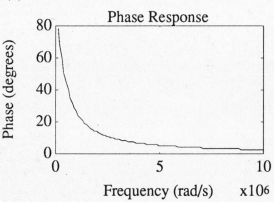

(c)

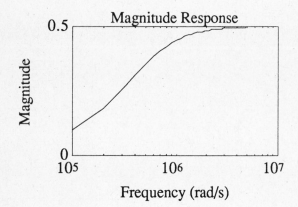

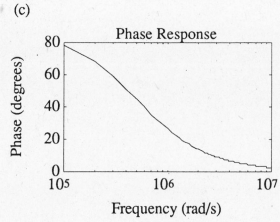

(d)

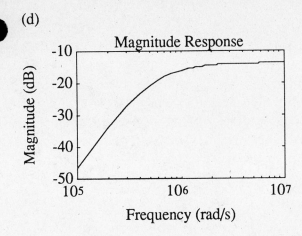

**5.5** Compute and plot the frequency response $\dfrac{V_3(j\omega)}{V_S(j\omega)}$ for the circuit of Figure P6.9.

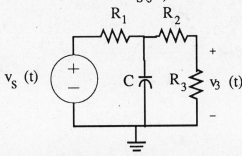

$R_1 = R_2 = 500\Omega \quad R_3 = 1000\Omega \quad C = 0.27\mu F$

**Figure P6.9**

**Solution:**

(a) Applying KVL at the node between $R_1$ and $R_2$, we have

$$\frac{V - V_S}{500} + \frac{V}{1/j\omega C} + \frac{V}{1500} = 0$$

$$V(3 + 1 + j\omega C1500) = V_S(3)$$

$$V = \frac{V_S(3)}{4 + j\omega C1500} = \frac{V_S(3/4)}{1 + j\omega C375}$$

By the voltage divider rule, we have

$$V_3 = V\frac{R_3}{R_2 + R_3}$$

$$\frac{V_3(\omega)}{V_S(\omega)} = \frac{3/4}{1 + j\omega C375}\left(\frac{1000}{1500}\right)$$

$$\left|\frac{V_{out}}{V_{in}}\right| = \frac{0.5}{\sqrt{1 + \frac{\omega^2}{(9876)^2}}}$$

$$\phi(\omega) = -\arctan\left(\frac{\omega}{9876}\right)$$

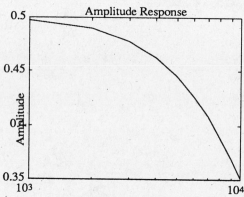

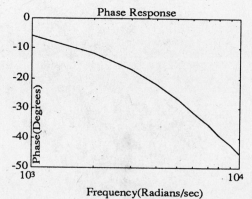

**5.6**  Determine the component of the output voltage, $v_C(t)$, in the circuit of Figure P6.10 due to:

a) $v_{S1}(t)$

b) $v_{S2}(t)$

c) Define the voltage attenuation as $\dfrac{|V_C(j\omega)|}{|V_S(j\omega)|}$ Find the attenuation of <u>each</u> of the     sources

for this circuit.  Also compute the attenuation in dB.

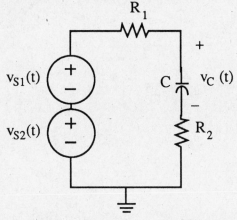

$$R_1 = R_2 = 1000\Omega \quad C = 1\mu F \quad V_{S1} = 10V \cos(500t) \quad V_{S2} = 10V \cos(5000t)$$

**Figure P6.10**

**Solution:**

(a) In general:

$$V_{out}(\omega) = V_S(\omega) \frac{1/j\omega C}{R1 + R2 + 1/j\omega C}$$

$$= V_S(\omega) \frac{1}{j\omega C(R_1 + R_2) + 1}$$

$$= V_S(\omega) \frac{1}{1 + j\omega/500}$$

The voltage due to $V_{S1}$ is:

$$V_C(\omega)|_{\omega = 500} = \frac{10}{1 + j} = 7.07\angle\text{-}45° \text{ V}$$

Therefore,

$$\boxed{v_C(t)|_{VS1} = 7.07 \cos(500t - 45°) \text{ V}}$$

(b)

$$V_C(\omega)|_{\omega = 5000} = \frac{10}{1 + j10} = 0.995\angle\text{-}84.3° \text{ V}$$

$$\boxed{v_C(t)|_{VS2} = 0.995 \cos(5000t - 84.3°) \text{ V}}$$

(c)  $v_C(t) = v_C(t)|_{vs1(t)} + v_C(t)|_{vs2(t)}$

$$v_C(t) = 7.07 \cos(500t - 45°) + 0.995 \cos(5000t - 84.3°) \text{ V}$$

Attenuation for $v_{S1}(t) = 0.707$

Attenuation for $v_{S2}(t) = 0.0995$

Attenuation in dB for $v_{S1}(t) = -3$ dB

Attenuation in dB for $v_{S2}(t) = -20$ dB

---

**5.7**  a)  Determine the frequency response, $\dfrac{V_o(j\omega)}{V_{in}(j\omega)}$, for the circuit of Figure 6.16 where

$R = 10$ k$\Omega$,  $C = 100$ µF and L $= 0.1$ H.

   b)  What is the center frequency of this band-pass filter circuit?

   c)  Plot the frequency response of this circuit on semilog paper in units of dB.

   d)  Define the *3 dB bandwidth* of the filter as the frequency span between points of 3dB attenuation.  Determine the 3 dB bandwidth of the filter analytically and from the plot of part c).

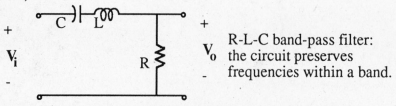

**Figure 6.16  R-L-C  Band-pass  filter**

**Solution:**

$$\left|\frac{V_o}{V_i}\right| = \frac{\omega RC}{\sqrt{(1 - \omega^2 LC)^2 + (\omega RC)^2}}$$

(a)   $$\left|\frac{V_o}{V_i}\right| = \frac{\omega}{\sqrt{(1 - 10^{-5}\omega^2)^2 + \omega^2}}$$

(b)   $$\omega_{center} \approx \frac{1}{\sqrt{LC}} = 316.2 \text{ rad/s}$$

(c)

(d)  Since an attenuation of -3 dB corresponds to a reduction by a factor of $1/\sqrt{2}$, we can find the desired frequency by solving the following equation:

$$\frac{1}{\sqrt{2}} = \frac{\omega}{\sqrt{(1 - 10^{-5}\omega^2)^2 + \omega^2}}$$

Solving for $\omega$, we obtain the quadratic equation

$$\omega^4 (LC)^2 - \omega^2 (2LC + (RC)^2) + 1 = 0$$

which can be solved to obtain $\omega_1 = 1$ rad/s and $\omega_2 = 100,000$ rad/s.  Therefore, the 3-dB bandwidth of the filter is $\omega_2 - \omega_1 \approx 100,000$ rad/s which is confirmed by the graph.

---

**5.8**   a)  Determine the frequency response, $\dfrac{V_o(j\omega)}{V_{in}(j\omega)}$, for the circuit of Figure 6.16 where

$R = 5\,\Omega$,  $C = 20\,\mu F$ and $L = 1$ H.

   b)  What is the center frequency of this band-pass filter circuit?

   c)  Plot the frequency response of this circuit on semilog paper in units of dB.

d) Define the *3 dB bandwidth* of the filter as the frequency span between points of 3dB attenuation. Determine the 3 dB bandwidth of the filter analytically and from the plot of part c).

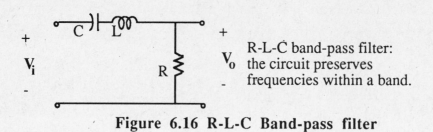

R-L-C band-pass filter:
the circuit preserves
frequencies within a band.

**Figure 6.16 R-L-C Band-pass filter**

**Solution:**

$$\left|\frac{V_o}{V_i}\right| = \frac{\omega RC}{\sqrt{(1 - \omega^2 LC)^2 + (\omega RC)^2}}$$

a) $$\left|\frac{V_o}{V_i}\right| = \frac{10^{-4}\omega}{\sqrt{(1 - 20(10^{-6})\omega^2)^2 + (10^{-4}\omega)^2}}$$

b) $$\omega_{center} = \frac{1}{\sqrt{LC}} = 223.6 \text{ rad/s}$$

c)

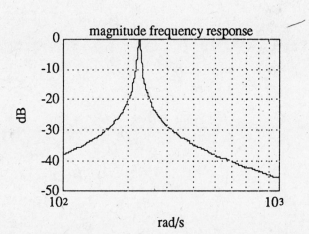

d) Since an attenuation of -3 dB corresponds to a reduction by a factor of $1/\sqrt{2}$, we can find the desired frequency by solving the following equation:

$$\frac{1}{\sqrt{2}} = \frac{\omega RC}{\sqrt{(1 - \omega^2 LC)^2 + (\omega RC)^2}}$$

solving for $\omega$, we obtain $\omega_1 = 221.1$ rad/s and $\omega_2 = 226.1$ rad/s. Therefore, the 3-dB bandwidth is $\omega_2 - \omega_1 = 5$ rad/s, which is confirmed by the graph.

5.15

**5.9** A phonograph driving an audio amplifier might be modelled as shown in the circuit of Figure P6.13:

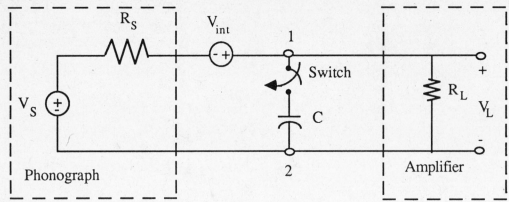

**Figure P6.13**

A "by-pass capacitor" is placed between terminals 1 and 2 to reduce pickup from a nearby broadcast station represented in the drawing above by the interference voltage, $V_{int}$.

a) If $V_{L0}$ is the voltage at the amplifier with the switch open and $V_L$ is the same voltage with the switch closed, find $\dfrac{V_L}{V_{L0}}$ in terms of $\omega$ if $R_S = R_L = 10^4 \Omega$ and $C = 10^{-9}F$

b) If the broadcast station's frequency is $10^7$ rad/sec, find the magnitude of $\dfrac{V_L}{V_{L0}}$.

c) Find $\left| \dfrac{V_L}{V_{L0}} \right|$ if the broadcast station's frequency is $10^5$ rad/sec

**Solution:**

(a) $V_{LO} = (V_{int} + V_S)(\dfrac{R_L}{R_S + R_L})$

$V_L = (V_{int} + V_S)\{\dfrac{R_L}{R_S + R_L + j\omega CR_LR_S}\}$

$$\dfrac{V_L}{V_{LO}} = \dfrac{\dfrac{R_L}{R_S + R_L + j\omega CR_LR_S}}{\dfrac{R_L}{R_S + R_L}}$$

$$= \dfrac{1}{1 + j\omega\dfrac{CR_LR_S}{R_S + R_L}}$$

$$\dfrac{V_L}{V_{LO}} = \dfrac{1}{1 + j\omega(5\times10^{-6})}$$

(b) $\omega = 10^7$ $\boxed{|\dfrac{V_L}{V_{LO}}| = 0.02}$

(c) $\omega = 10^5$ $\boxed{|\dfrac{V_L}{V_{LO}}| = 0.894}$

---

**5.10** For the circuit shown in Figure P6.21, assume switch $S_1$ is always open and switch $S_2$ closes at t = 0.

   (a) Find the inductor current at $t = 0^+$

   (b) Find the time constant, $\tau$, for $t \geq 0$.

   (c) Find an expression for $i_L(t)$ and sketch the function

   (d) Find $i_L(t)$ for each of the following values of t: $0$, $\tau$, $2\tau$, $5\tau$, $10\tau$

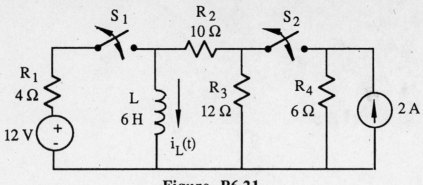

**Figure  P6.21**

**Solution:**

a) $i_L(0^-) = i_L(0^+) = 0$ A

b) $\tau = 3/7$ s

c)

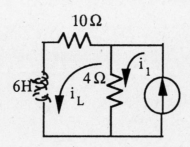

Using mesh current analysis:

$$\frac{di_L}{dt} + \frac{7}{3} i_L = \frac{4}{3}$$

Solving the differential equation:

$$i_L(t) = k_1 e^{-7/3 \, t} + \frac{4}{7} A \quad t > 0$$

$i_L(0^+) = 0.$

Thus,

$$i_L(t) = -\frac{4}{7} e^{-7/3\, t} + \frac{4}{7} \text{ A} \qquad t > 0$$

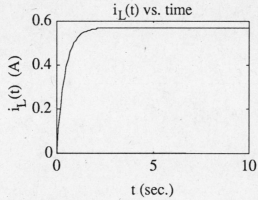

$i_L(t)$ vs. time

d)  $i_L(0) = 0$ A $\qquad$ $i_L(\tau) = 0.36$ A;

$i_L(2\tau) = 0.49$ A $\qquad$ $i_L(5\tau) = 0.57$ A;

$i_L(10\tau) = 0.57$ A

---

**5.11** For the circuit shown in Figure P6.21, assume that switch $S_1$ is always open; switch $S_2$ has been closed for a long time, and opens at $t = 0$.

(a) Find the inductor current at $t = 0^+$

(b) Find the time constant $\tau$, for $t \geq 0$.

(c) Find an expression for $i_L(t)$ and sketch the function.

(d) Find $i_L(t)$ for each of the following values of t: $0, \tau, 2\tau, 5\tau, 10\tau$.

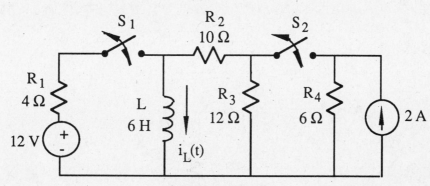

**Figure P6.21**

## Solution:

a)  $i_L(0^-) = i_L(0^+) = 4/7$ A

b)  $\tau = 3/11$ s

c)

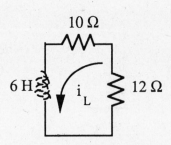

Applying KVL:

$$i_L(t) = k_1\, e^{-11/3t}\ \text{A}$$

Solving for the initial condition:

$$i_L(t) = 4/7\, e^{-11/3t}\ \text{A}$$

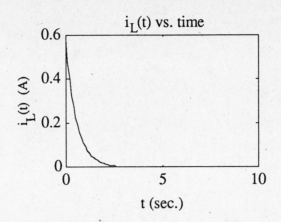

d)    $i_L(0) = 4/7$ A    $i_L(\tau) = 0.21$ A;

$i_L(2\tau) = 0.08$ A    $i_L(5\tau) = 0.004$ A;

$i_L(10\tau) = 2.6 \times 10^{-5}$ A

---

**5.12** For the circuit shown in Figure P6.21, assume switch $S_1$ is always open; switch $S_2$ has been closed for a long time and opens at $t = 0$. At $t = t_1 = 3\tau$, switch $S_2$ closes again.

(a) Find the inductor current at $t = 0^+$

(b) Find the expression for $i_L(t)$ for $t > 0$ and sketch the function

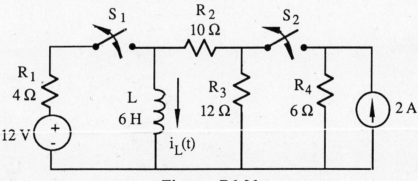

**Figure  P6.21**

**Solution:**

a) From 6.29, we have

$$i_L(3\tau) = 0.03 \text{ A}$$

Therefore,

$$i_L(0^-) = i_L(0^+) = 0.03 \text{ A}$$

c)

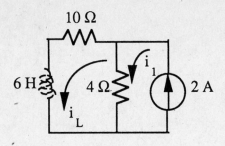

Using mesh current analysis:

$$\frac{di_L}{dt} + \frac{7}{3}i_L = \frac{4}{3}$$

Solving the differential equation:

$$i_L(t) = k_1 e^{-7/3\,t} + \frac{4}{7}\,A \qquad t > 0$$

$$i_L(0^+) = 0.03\ A.$$

Thus,

$$i_L(t) = -0.54e^{-7/3\,t} + 0.57A \qquad t > 0$$

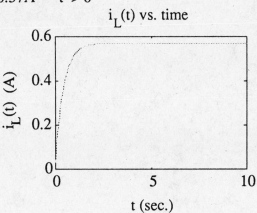

$i_L(t)$ vs. time

5.13 Assume both switches $S_1$ and $S_2$ in Figure P6.21 close at $t = 0$.

(a) Find the inductor current at $t = 0^+$

(b) Find the time constant $\tau$, for $t \geq 0$.

(c) Find an expression $i_L(t)$ and sketch the function.

(d) Find $i_L(t)$ for each of the following values of t: $0, \tau, 2\tau, 5\tau, 10\tau$.

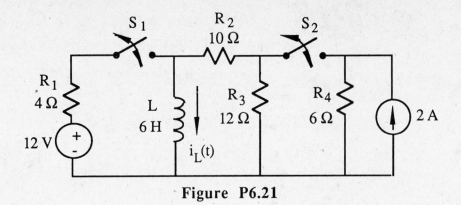

**Figure P6.21**

**Solution:**

a) $i_L(0^-) = i_L(0^+) = 0A$

b) $\tau = 27/14$ s

c)

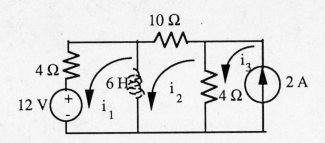

Applying mesh current analysis:

$$i_L(t) = k_1 e^{-14/27\, t} + \frac{50}{14}A \qquad t > 0$$

$$i_L(0^+) = 0 \text{ A}.$$

Thus,

$$i_L(t) = -50/14\, e^{-14/27\, t} + 50/14 \text{ A} \qquad t > 0$$

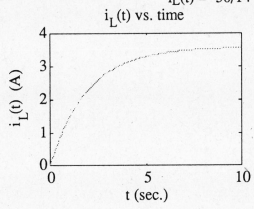

$i_L(t)$ vs. time

d) $i_L(0) = 0$ A; $\qquad i_L(\tau) = 2.26$ A;

$i_L(2\tau) = 3.09$ A; $\qquad i_L(5\tau) = 3.55$ A;

$i_L(10\tau) = 3.57$ A

**5.14** Assume both switches $S_1$ and $S_2$ in Figure P6.21 have been closed for a long time, and that switch $S_1$ opens at $t = 0^+$.

  (a) Find the inductor current at $t = 0^+$.

  (b) Find an expression for $i_L(t)$ and sketch the function.

  (c) Find $i_L(t)$ for each of the following values of t: $0, \tau, 2\tau, 5\tau, 10\tau$.

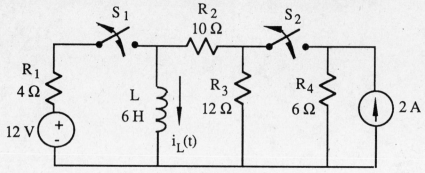

**Figure P6.21**

**Solution:**

a)

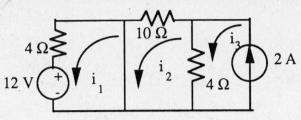

Applying mesh current analysis:

$$i_L(0^-) = i_L(0^+) = i_2 - i_1 = 50/14 \text{ A}$$

b)

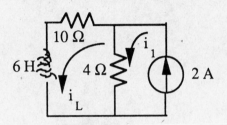

Using mesh current analysis:

$$\frac{di_L}{dt} + \frac{7}{3} i_L = \frac{4}{3}$$

Solving the differential equation:

$$i_L(t) = k_1 e^{-7/3\,t} + \frac{4}{7} \text{ A} \qquad t > 0$$

$$i_L(0^+) = 25/7 \text{ A}.$$

Thus,

$$i_L(t) = 3 e^{-7/3\,t} + 4/7 \text{ A} \qquad t > 0$$

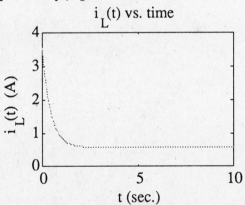

c)   $i_L(0) = 25/7$ A;     $i_L(\tau) = 1.68$ A;

   $i_L(2\tau) = 0.98$ A;     $i_L(5\tau) = 0.59$ A;

   $i_L(10\tau) = 0.57$ A

**5.15**  The circuit of Figure P6.26 is used as a variable delay in a burglar alarm.  The alarm is a siren with internal resistance of 1KΩ.  The alarm will not sound until the current $i_L$ exceeds 100μA.  Find the range of the variable resistor, R, such that the delay is between 1 and 2 seconds.

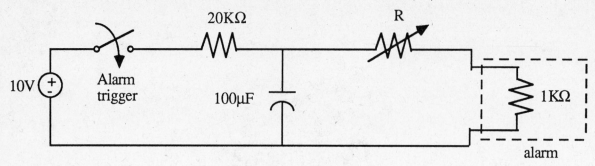

**Figure  P6.26**

**Solution:**

When $i_L = 100 \, \mu A$,

$$v_C = (1000 + R) \times 100 \times 10^{-6}$$
$$v_C = \frac{1000 + R}{10000}$$

A general expression for the capacitor voltage is:

$$v_C(t) = \frac{R + 1000}{20000 + 1000 + R} \times 10 \times (1 - e^{-t/\tau})$$

At the time the alarm sounds, we have

$$v_C = \frac{1000 + R}{10000}$$

$$= \frac{R + 1000}{21000 + R} \times 10 \times (1 - e^{-t/\tau})$$

$$1 - \left(\frac{21000 + R}{100000}\right) = e^{-t/\tau}$$

The time constant is given by the expression

$$\tau = 100 \, \mu F[(1000 + R) \parallel 20000]$$

Substituting the expression for the time constant into the above equation we have one equation in one unknown, R. This is a transcendental equation, and can be solved by iteration or by graphical analysis. the solution is that R must be between 33,580 $\Omega$ and 53,510 $\Omega$ or

$$33,580 \ \Omega \leq R \leq 53,510 \ \Omega$$

**5.16** Find the voltage across $C_1$ in the circuit of Figure P6.27 for $t>0$. $C_1 = 5\mu F$, $C_2 = 10\mu F$. Assume the capacitors are initially uncharged.

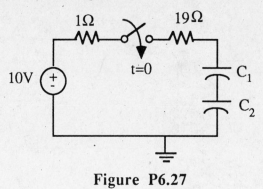

**Figure P6.27**

**Solution:**

$$C_{eq} = \frac{C_1 C_2}{C_1 + C_2} = \frac{50}{15}\,\mu F = \frac{10}{3}\,\mu F$$

$$\tau = R_{eq}C_{eq} = (20)\left(\frac{10}{3}\right)\mu s = 66.7\,\mu s$$

$$i(0) = \frac{10V}{(1+19)}\,\Omega = 0.5A$$

$$i(t) = i(0)e^{-t/\tau} = 0.5\,e^{-t/66.7\mu S}$$

$$v_C(t) = \frac{1}{C}\int_{t_0}^{t} i(\lambda)d\lambda$$

$$v_{c1}(t) = \frac{1}{5\mu F}\int_{0}^{t} 0.5e^{-\lambda/66.7\mu S}d\lambda$$

$$= \frac{66.7\mu S}{5\mu F}(0.5)\left[e^{-\lambda/66.7\mu S}\right]_{0}^{t}$$

$$= -6.67\left[e^{-t/66.7\mu S} - 1\right]$$

$$\boxed{v_{C1}(t) = 6.67 - 6.67e^{-t/66.7\mu s}\;V}$$

---

**5.17**  The circuit of Figure P6.32 has a switch which connects and disconnects a battery to and from the rest of the circuit.  For $t < 0$ the switch has been open for a very long time.  At $t = 0$ the switch closes and then at $t = 50\ mS$ the switch opens again.

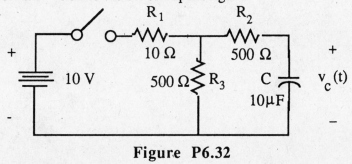

**Figure  P6.32**

a)  Determine the capacitor voltage as a function of time (don't forget that there are   three periods of time that must be considered here)

b)  Plot the capacitor voltage from t=0 to t=100 mS.

**Solution:**

The Thèvenin equivalent seen by the capacitor with the switch closed is:

$$R_T = 10 \parallel 500 + 500 = 510 \ \Omega$$

$$V_T = \frac{500}{500 + 10} \ 10 = 9.8 \ V$$

For $t < 0$, $v_C(0^-) = 0$; thus, applying KVL:

$$V_T = i(t) \ R_T + \frac{1}{C} \int_0^t i(t)dt + v_C(0^-)$$

Differentiating

$$\frac{di(t)}{dt} R_T = -\frac{i(t)}{C}$$

$$\int_0^t \frac{di(\tau)}{i(\tau)} d\tau = \int_0^t - \frac{1}{R_TC} d\tau$$

$$\ln[i(t)]|_{t,,0} = \frac{-1}{R_TC}(t - 0)$$

$$i(t) = i(0)e^{-t/RTC}$$

$$v_C(t) = \frac{1}{C} \int_0^t i(t)\, dt = \frac{1}{C} \int_0^t i(0)\, e^{-t/RTC}\, dt$$

$$= -i(0)R_T (e^{-t/RTC} - 1) = V_T(1 - e^{-t/RTC})$$

$$= 9.8(1 - e^{-t/RTC})$$

For $0 < t < 50$ ms,

$$v_C(t) = 9.8 (1 - e^{t/5.1ms})$$

Now at $t > 50$ ms, we find that the switch opens and we are now concerned with the following circuit:

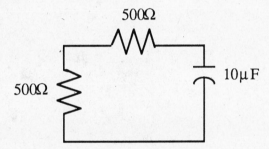

the capacitor's voltage when the switch opens is

$$v_C(50\text{ ms}) = V_T(1 - e^{50/51}) = 9.8 \text{ V}$$

Now the path for the capacitor discharge is through the 500-$\Omega$ resistors. If we write KVL and solve the differential equation again, we find that

$$V_C(t) = 9.8\, e^{-100(t - .05)} \text{ V}$$

where t is in seconds.

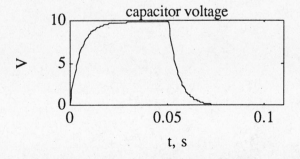

**5.18** The ideal current source in the circuit of Figure P6.33 switches between several current levels as shown in the graph.

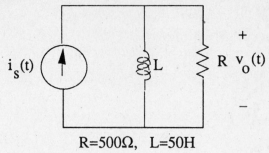

R=500Ω,  L=50H

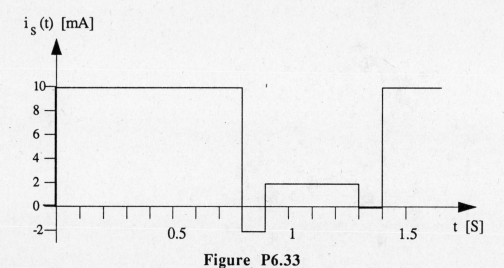

**Figure P6.33**

Determine and sketch the voltage across the inductor $v_L(t)$ for t between 0 and 2 seconds. Note that there are six different time segments to consider for this problem. When you consider each time interval,you must, in effect, switch your origin, that is, if we assume $t_0=0$ for the origin for one interval and the next interval begins at $t_1$, you might write the exponential term in the solution as $e^{-(t-t_1)/\tau}$ for the second interval. You may assume that the current source has been zero for a very long time before t = 0.

**Solution:**

Applying KCL :

$$\frac{V}{R} + \frac{1}{L} \int v_o \, dt = i_S(t)$$

Differentiating, we have

$$\frac{dv_o}{dt} = -\frac{R}{L} v_o$$

$$\log_e\left(\frac{v_o(t)}{v_o(t_0)}\right) = -\frac{R}{L}(t - t_0)$$

$$v_o(t) = v_o(t_0)e^{(t - t_0)}$$

$$v_o(t) = v_o(t_0)e^{-10(t - t_0)}$$

region I:     $t < 0$    $v(t) = 0$

region II:    $0 < t < 0.8$ s

Since the current through an inductor cannot change instantaneously,

$$v_{oII}(t) = 5\, e^{-t/0.1}\ V$$

region III:    $0.8 < t < 0.9$ sec

$$v_{oII}^{-}(t = 0.8) = 0\ \ V$$
$$I_L(t = 0.8) = 10\ \ mA$$
$$i_R(t = 0.8^+) = -\,(10 + 2) = -\,12\ \ mA$$
$$v_o(t_0 = 0.8^+) = -\,12 \times 10^{-3} \times 500 = -\,6\ \ V$$
$$v_{oIII}(t) = -\,6\, e^{-(t - 0.8)/0.1}\ V$$

region IV:    $0.9 < t < 1.3$ sec

$$v_{oIII}^{-}(t = 0.9) = -\,2.2\ \ V$$
$$i_R(t = 0.9^-) = -\,4.4\ \ mA$$

KCL:

$$i_L(t = 0.9^-) = i_L(t = 0.9^+) = 2.4\ \ mA$$

At $t = 0.9^+$, we have

$$i_R(t) = 2 - 2.4 = -\,0.4\ \ mA$$
$$v_o(t_0 = 0.9^+) = 0.2\ \ V$$
$$v_{oIV}(t) = -\,0.2\, e^{-(t - 0.9)/0.1}\ \ V$$

region V:    $1.3 < t < 1.4$ sec

$$v_{oV}^{-}(t = 1.3) = -\,3.66\ mV$$

$$i_R(t = 1.3^-) = -\,7.32\ \mu A = 0\ \ A$$

$$i_L(t = 1.3^-) = i_L(t = 1.3^+) = 2.0\ \ mA$$
$$i_R(t = 1.3^+) = -\,2\ \ mA$$
$$v_{oV}(t_0 = 1.3^+) = -\,1\ \ V$$
$$v_{oV}(t) = -\,e^{-(t - 1.3)/0.1}\ \ V$$

Sketch:

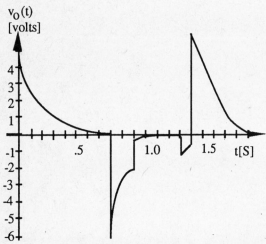

# Section 6: Semiconductors and Diodes

**6.1**   Consider the circuit of Figure P7.1. For $v_B = 15V$, determine whether the diode is conducting or not. Assume the diode is an ideal diode.

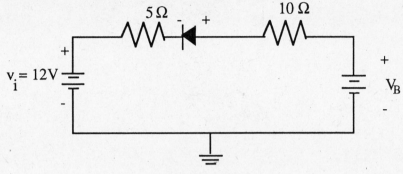

**Figure P7.1**

**Solution:**

Assuming the diode is conducting, the current is found to be

$$I = \frac{V_B - V_i}{5 + 10} = \frac{3}{15} = 0.2 \text{ A}$$

Since the result is consistent with the assumption, the diode is on.

---

**6.2**   Consider the circuit of Figure P7.2. Determine whether the diode is conducting or not. Assume the diode is an ideal diode.

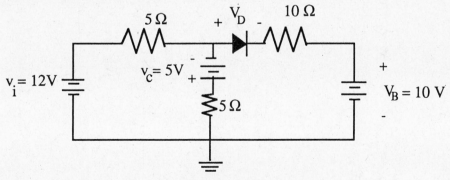

**Figure P7.2**

**Solution:**

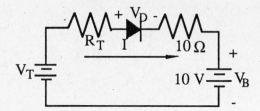

The Thèvenin equivalent resistance of the circuit to the left of the diode is $R_T = 2.5$ $\Omega$. The Thèvenin equivalent voltage is:

$$V_T = (\frac{V_i + V_C}{10}) \times 5 - V_C = 3.5 \ V$$

Assuming the diode is conducting,

$$I = \frac{3.5 - 10}{12.5} = -0.52 \ A$$

This contradicts the assumption. Thus the diode is off.

---

**6.3** Consider the circuit of Figure P7.2. Determine whether the diode is conducting or not for $v_i = 12V$, $v_B = 15V$, $v_c = 5V$. Assume the diode is an ideal diode.

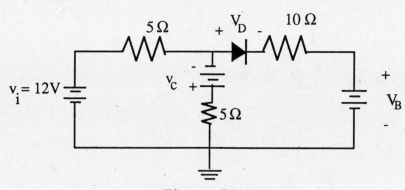

**Figure P7.2**

## Solution:

Assuming the diode is conducting, the current I is

$$I = \frac{3.5 - 15}{12.5} = -0.92 \text{ A}$$

The initial assumption cannot hold, since the result is a negative current; thus the diode must be off.

---

**6.4**   Consider the circuit of Figure P7.2. Determine whether the diode is conducting or not for $v_i = 12V$, $v_B = 15V$, $v_c = 15V$. Assume the diode is an ideal diode.

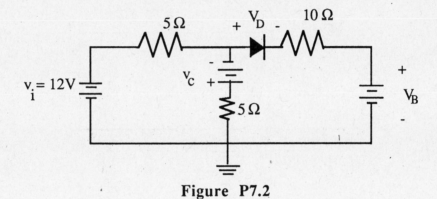

**Figure P7.2**

**Solution:**

The Thèvenin equivalent voltage is:

$$V_T = (\frac{V_i + V_C}{10}) \times 5 - V_C$$

$$= 2.7 \times 5 - 15 = -1.5 \text{ V}$$

Assuming the diode is conducting, the diode current I is

$$I = \frac{-1.5 - 10}{12.5} = -0.92 \text{ A}$$

The initial assumption cannot hold, since the result indicates a negative diode current. Thus, the diode is off.

---

**6.5** Consider the circuit of Figure P7.2. Determine whether the diode is conducting or not for $v_i = 12V$, $v_B = 15V$, $v_c = 10V$. Assume the diode is an ideal diode.

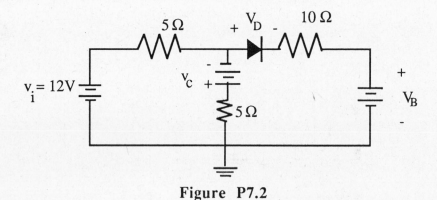

**Figure P7.2**

**Solution:**

The Thèvenin equivalent voltage is:

$$V_T = (\frac{V_i + V_C}{10}) \times 5 - V_C$$

$$= 2.2 \times 5 - 10 = 1 \text{ V}$$

Assuming the diode is conducting, the diode current I is

$$I = \frac{1 - 10}{12.5} = -0.72 \text{ A}$$

The initial assumption cannot hold, since the result indicates a negative diode current. Thus, the diode is off.

---

**6.6** For the circuit of Figure P7.5, assume $D_1$ and $D_2$ are ideal diodes.

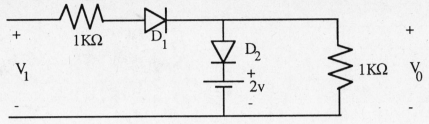

**Figure P7.5**

a. For what range of values of $V_1$ is diode $D_1$ forward biased?

b. For what range of values of $V_1$ is diode $D_2$ forward biased?

c. What is $V_0$ when $V_1 = 3$ V?

**Solution:**

(b) If $D_2$ is forward biased, the voltage at the node to the right of $D_1$ is 2 V. By the voltage divider rule we have

$$2 = V_1\left(\frac{1}{1 + 1}\right)$$

Therefore, $V_1 \geq 4$ V, in order to forward bias $D_2$.

(a) (1) Assume $D_2$ is open. Then, if $V_1 \geq 0$, $D_1$ is forward biased with current $V_1/2000$ A.

(2) Assume $D_2$ is conducting; in this case, the voltage at the node to the right of $D_1$ is 2 V. For $D_1$ to be forward-biased, the input voltage must therefore be

$$V_1 \geq 2 \text{ V}$$

This condition is satisfied, since $V_1 \geq 4$ V for $D_2$ to conduct. Therefore, $D_1$ is forward-biased for any $V_1 \geq 0$.

(c) From (a) and (b), we know that $D_1$ is forward biased, and $D_2$ is cut off; by the voltage divider rule, we find

$$V_0 = 1.5 \text{ V}$$

**6.7** A Zener diode is used in a switching circuit as shown in Figure P7.11.

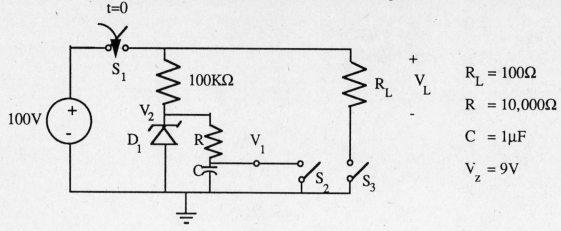

$R_L = 100\Omega$

$R = 10,000\Omega$

$C = 1\mu F$

$V_z = 9V$

**Figure  P7.11**

When $V_1$ reaches 7.2 V both switches $S_2$ and $S_3$ close simultaneously and stay closed for 3.9 ms, then they both open.

    a) Find the frequency for the voltage across the load resistor, $R_L$.

    b) Find the power dissipated by the resistor $R_L$.

HINT:  At $t = 0$, $S_1$ closes and $S_2$ and $S_3$ are open.  Find the waveforms for $V_1$ and $V_L$ and proceed from there.

**Solution:**

At $t = 0$, $S_1$ closes and $V_2$ is the Zener voltage, that is

$$V_2 = V_Z = 9 \text{ V}. \quad V_C(0) = 0 \text{ V}$$

The time constant, $\tau$, of the RC circuit is

$$\tau = RC = 10^4 \times 1 \times 10^{-6} = 10 \text{ ms}$$

The voltage $V_1$ is

$$V_1 = v_C(t) = v_C(\infty) - (v_C(\infty) - v_C(0))e^{-t/\tau} \text{ therefore}$$

$$V_1 = v_C(t) = 9 - 9e^{-t/\tau}$$

(a) When $V_1$ reaches 7.2 V, $S_3$ will close and $V_L$ will be 100 V for 3.9 ms.

$$v_C(t_1) = 7.2 = 9 - 9e^{-t1/\tau}$$

$$t_1 = -\tau \times \ln\frac{1.8}{9} = 16.1 \text{ ms}$$

The frequency for the voltage $V_L$ is

$$f_L = \frac{1}{T} = \frac{10^3}{16.1 + 3.9} = 50 \text{ Hz}$$

(b) The waveform of $v_L$ is shown below

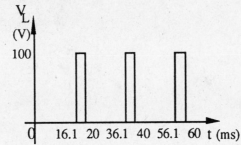

The rms value of voltage $V_L$ is

$$V_L{}^2{}_{rms} = \frac{1}{T} \int_0^T V_L{}^2 \, dt = \frac{3.9 \times 10^4}{20} = 1950$$

$$V_{Lrms} = 44.16 \text{ V}$$

The power dissipated in $R_L$ is

$$P_L = \frac{V_L{}^2{}_{rms}}{R_L} = 19.5 \text{ W}$$

---

**6.8** You have purchased an inexpensive diode which has the i-v characteristics shown in Figure P7.18.

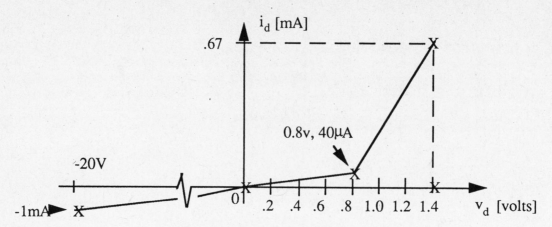

**Figure  P7.18**

Determine the picewise linear model using the form shown in Figure P7.17.

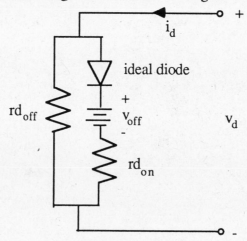

**Figure  P7.17**

**Solution:**

From the i-v characteristics, we have

$$r_{Don} = \frac{\Delta v_D}{\Delta i_D}$$

$$= \frac{1.4 - 0.8}{0.67 \times 10^{-3} - 40 \times 10^{-6}} = 952 \ \Omega$$

$$V_\gamma = 0.8 \ V$$

$$r_{Doff} = \frac{20}{1 \times 10^{-3}} = 20 \ k\Omega$$

---

**6.9** Sketch the output waveform for the circuit of Figure P7.21. Assume ideal diode characteristics, $v_s(t) = 10 \sin (2000\pi t)$.

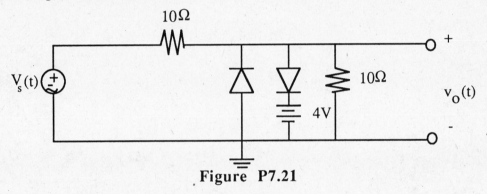

Figure P7.21

**Solution:**

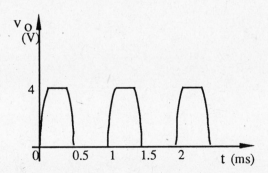

---

**6.10**  Find the output waveform for the circuit of Figure P7.23.

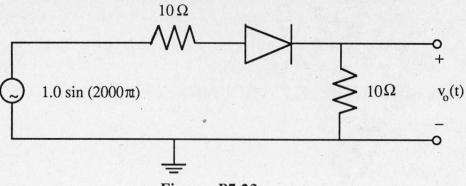

Figure  P7.23

**Solution:**

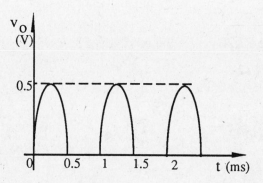

**6.11**  Find the output waveform for the circuit of Figure P7.25.

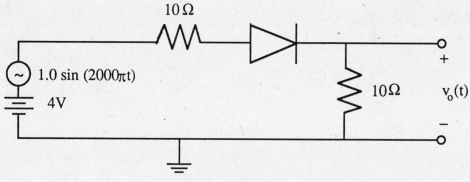

Figure  P7.25

**Solution:**

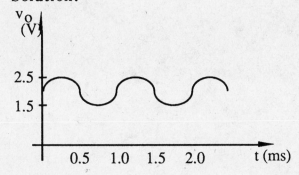

**6.12** The circuit of Figure P7.26 forces the diode to operate at a certain voltage and current. If the diode has the given i-v characteristics find the operating point of the diode.

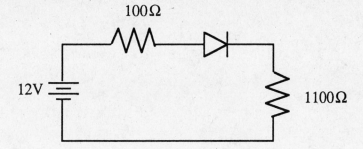

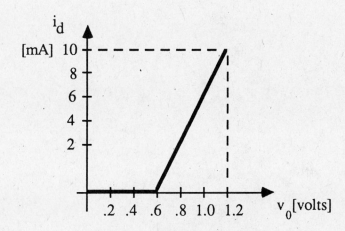

**Figure P7.26**

6.11

**Solution:**

$$r_{Don} = \frac{1.2 - 0.6}{10 \text{ mA}} = 60 \ \Omega$$

$$V_\gamma = 0.6 \ V$$

$$r_{Doff} = \infty$$

The operating point is

$$i_{DQ} = \frac{12 - 0.6}{1200 + 60} = 9.05 \ \text{mA}$$

$$v_{DQ} = 60 i_{DQ} + 0.6 = 1.143 \ V$$

---

**6.13**  For the voltage limiter circuit of Figure P7.32, plot the voltage across $R_L$ vs. $v_S$ for $-20 < v_S < 20$.

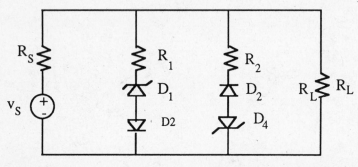

$R_S = 10\Omega$

$R_1 = 1\Omega$

$R_2 = 10$

$R_L = 40\Omega$

$V_{2_1} = 10V$

$V_{2_4} = 5V$

Treat all diodes as ideal diodes

**Figure  P7.32**

Three cases need to be considered:

(a)  When $v_L > 10$ V, $D_1$ will conduct

(b)  When $v_L < -5$ V, $D_4$ will conduct

(c)  When $-5 < v_L < 10$, all 4 diodes will be cut off.

## Solution:

Three cases need to be considered:

(a)  When $v_L > 10$ V, $D_1$ will conduct

(b)  When $v_L < -5$ V, $D_4$ will conduct

(c)  When $-5 < v_L < 10$, all 4 diodes will be cut off.

For case (a), the circuit is shown below.

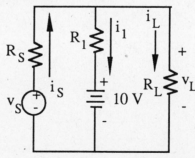

By KCL, $i_S = i_1 + i_L$, and

$$\frac{v_S - v_L}{10} = \frac{v_L - 10}{1} + \frac{v_L}{40}$$

Therefore,   $v_L = \dfrac{4v_S + 400}{45}$ V

$$v_L > 10 \text{ V} \implies v_S > 12.5 \text{ V}$$

For case (b), the circuit is shown below.

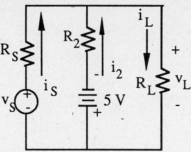

By KCL, $i_S + i_2 = i_L$, and

$$\frac{v_S - v_L}{10} + \frac{-5 - v_L}{10} = \frac{v_L}{40}$$

Therefore,    $v_L = \frac{4v_S - 20}{9}$  V

$$v_L < -5 \text{ V} \implies v_S < -6.25 \text{ V}$$

For case (c), we have $v_L = \frac{4}{5} v_S$ and

$$-6.25 \text{ V} < v_S < 12.5 \text{ V}$$

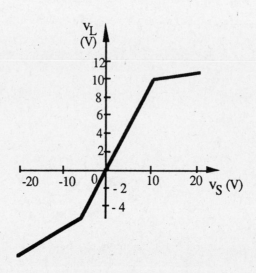

---

**6.14**    For the diode clipper of Figure 7.58, find the percentage of the source voltage which reaches the load if $R_L = 1$ k$\Omega$, $r_S = 100$ $\Omega$, and $r_D = 10$ $\Omega$. Assume the diode is conducting and use the circuit model of Figure 7.63.

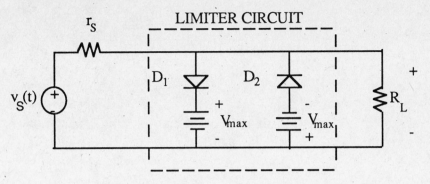

A two-sided diode voltage limiter

**Figure 7.58   Two-sided diode clipper**

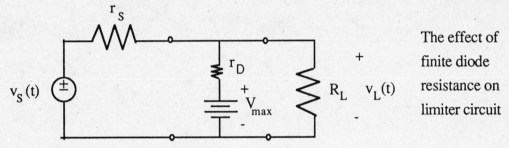

The effect of finite diode resistance on limiter circuit

**Figure 7.63   Circuit model for diode clipper (piece-wise linear diode)**

**Solution:**

By applying the superposition method, we have

$$v_L(t) = \frac{R_L \| r_S}{r_D + R_L \| r_S} V_{max} + \frac{r_D \| R_L}{r_S + r_D \| R_L} v_S(t)$$

Therefore, the percentage of the source voltage that reaches the load is

$$\frac{v_S(t)}{v_L(t)} \times 100 = \frac{r_D \| R_L}{r_S + r_D \| R_L} \times 100 = \frac{990}{109.90}$$

$$= 9.01\%$$

**6.15** For the diode clipper of Figure 7.58, find the percentage of the source voltage which reaches the load if $R_L = 10\ k\Omega$, $r_S = 500\ \Omega$ and $r_D = 50\ \Omega$. Assume the diode is conducting and use the circuit model of Figure 7.63.

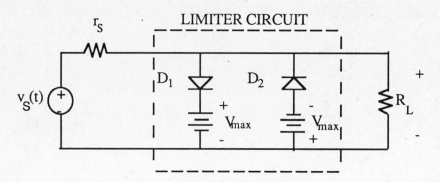

A two-sided diode voltage limiter

**Figure 7.58   Two-sided  diode  clipper**

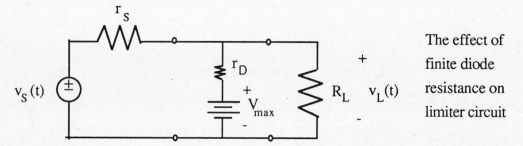

The effect of finite diode resistance on limiter circuit

**Figure 7.63   Circuit  model  for  diode  clipper  (piece-wise  linear  diode)**

**Solution:**

By applying the principle of superposition,

$$v_L(t) = \frac{R_L \| r_S}{r_D + R_L \| r_S} V_{max} + \frac{r_D \| R_L}{r_S + r_D \| R_L} v_S(t)$$

Therefore, the percentage of the source voltage that reaches the load is

$$\frac{v_S(t)}{v_L(t)} \times 100 = \frac{r_D \| R_L}{r_S + r_D \| R_L} \times 100$$

$$= \frac{4975}{500 + 49.75} = 9.05\%$$

## Section 7: Bipolar Junction Transistors

**7.1** For the circuit in Figure 8.11, the potentials of the three terminals are measured with respect to the ground: $V_1 = 6.3$ V, $V_2 = 5.5$ V, $V_3 = 6.0$ V. Determine the transistor operating state.

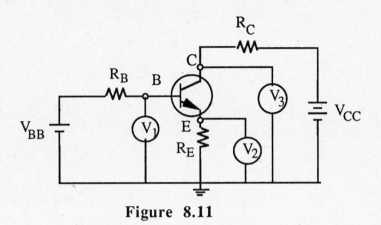

**Figure 8.11**

**Solution:**

The voltage at the BE junction is

$$V_{BE} = V_1 - V_2 = 6.3 - 5.5 = 0.8 \text{ V}$$

and the junction is forward biased

The voltage at the CB junction is

$V_{CB} = V_3 - V_1 = 6.0 - 6.3 = -0.3$ V and the junction is forward biased. Therefore, the transistor is operating in the saturation region.

---

**7.2** An *npn* transistor is operated in the active region with $I_B = 0.2$ mA, $I_C = 12$ mA, $V_{BE} = 0.7$ V and $V_{CE} = 14$ V, find

    (a) $I_C/I_E$

    (b) $V_{CB}/V_{BE}$

    (c) the total power dissipated by the transistor

**Solution:**

$\beta$ is found to be

$$\beta = \frac{I_C}{I_B} = 60$$

(a)  $I_E = I_C + I_B = \dfrac{\beta + 1}{\beta} I_C = 12.2 \ mA$

Therefore

$$\frac{I_C}{I_E} = 0.98$$

(b)  $V_{CB} = V_{CE} - V_{BE} = V_{CE} + V_{EB}$

$$= 14 - 0.7 = 13.3 \ V$$

Therefore  $\dfrac{V_{CB}}{V_{BE}} = 19$

(c)  The total power dissipated by the transistor is

$P \approx V_{CE}I_C = 14 \times 12 \times 10^{-3} = 0.168 \ W$

---

**7.3**  For the circuit given in Figure 8.11, the following readings are measured:

(a)  $V_1 = 1.62 \ V$, $V_2 = 1 \ V$, $V_3 = 1.2 \ V$

(b)  $V_1 = 3.2 \ V$, $V_2 = 2.5 \ V$, $V_3 = 7 \ V$

(c)  $V_1 = 0.2 \ V$, $V_2 = 0 \ V$

What is the operating state of the transistor?

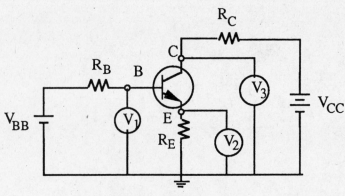

**Figure  8.11**

**Solution:**

(a) $V_{BE} = V_1 - V_2 = 1.62 - 1 = 0.62$ V, the BE junction is forward biased.

$V_{CB} = V_3 - V_1$

$= 1.2 - 1.62 = -0.42$ V

The CB junction is forward biased. The transistor is in the saturation region.

(b) $V_{BE} = V_1 - V_2 = 3.2 - 2.5 = 0.7$ V, the BE junction is forward biased.

$V_{CB} = V_3 - V_1 = 7 - 3.2 = 3.8$ V, the CB junction is reverse biased.

The transistor is in the active region.

(c) $V_{BE} = V_1 - V_2 = 0.2 - 0 = 0.2$ V, the BE junction voltage is smaller than $V_\gamma$,

therefore, it is reverse biased and the transistor is in the cut-off region.

---

**7.4** For the circuit shown in Figure 8.20, $V_{cc} = 20V$, $R_C = 5$ k$\Omega$, $R_E = 1$ k$\Omega$. Determine the operating state of the transistor, if

(a) $I_C = 1$ mA, $I_B = 0.2$ mA, $V_{BE} = 0.7$ V

(b) $I_C = 3.2$ mA, $I_B = 0.3$ mA, $V_{BE} = 0.8$ V

(c) $I_C = 3$ mA, $I_B = 1.5$ mA, $V_{BE} = 0.85$ V

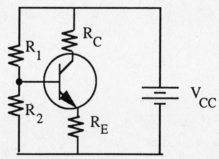

**Figure 8.20   Practical BJT self-bias DC circuit**

**Solution:**

(a) $V_C = V_{CC} - I_C R_C$

$= 20 - 1 \times 5 = 15$ V

$V_E = I_E R_E = (I_B + I_C) R_E$

$= 1.02 \times 1 = 1.02$ V

$V_{CB} = V_C - V_B = V_C - (V_{BE} + V_E)$

$= 15 - (0.7 + 1.02) = 13.28$ V

The transistor is in the active region.

(b) $V_C = V_{CC} - I_C R_C$

$= 20 - 3.2 \times 5 = 4$ V

$V_E = I_E R_E = (I_B + I_C) R_E$

$= 3.5 \times 1 = 3.5$ V

$V_{CB} = V_C - V_B = V_C - (V_{BE} + V_E)$

$= 4 - (0.8 + 3.5) = -30$ mV

The transistor is in the saturation region.

(c) $V_C = V_{CC} - I_C R_C = 20 - 3 \times 5 = 5$ V

$V_E = I_E R_E = (I_B + I_C) R_E$

$= 4.5 \times 1 = 4.5$ V

$V_{CB} = V_C - V_B = V_C - (V_{BE} + V_E)$

$= 5 - (0.85 + 4.5) = -350$ mV

The transistor is in the saturation region.

---

**7.5** For the following conditions, determine if each applies to active region operation of an *npn* transistor, a pnp transistor or both, or neither.

(a) $V_{EB} = 0.6$ V

(b) $I_C > I_E$

(c) $|I_C| > |I_E|$

(d) $I_E > I_B$

(e) $V_{CE} > V_{CB}$

**Solution:**

(a) pnp

(b) neither

(c) neither

(d) both

(e) both

---

**7.6**    Use the collector characteristics of the 2N3904 npn transistor ( $\beta = 150$ ) to determine the operating point ($I_{CQ}$, $V_{CEQ}$) of the transistor in the Figure P8.6.  Also determine $\beta$.

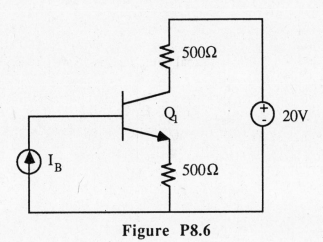

**Figure  P8.6**

a)  $I_B = 0.1$ mA

b)  $I_B = 0.19$ mA

c)  $I_B = 0.04$ mA

**Solution:**

From the data sheet, we can see that a typical value of $\beta$ is 150. The expression for the collector-emitter voltage is:

$$V_{CE} = V_{CC} - R_C I_C - R_E I_E$$

$$= 20 - 500 I_C - 500 \frac{\beta+1}{\beta} I_C$$

$$= 20 - 500 \beta I_B (1 + \frac{\beta+1}{\beta})$$

$$= 20 - 150{,}500 I_B$$

(a) $I_C = \beta I_B = 15$ mA, $V_{CE} = 4.95$ V

(b) $I_C = \beta I_B = 28.5$ mA, $V_{CE} = -8.595$ V

This means that the transistor is saturated, and the actual $\beta$ must be much less than 150. Assuming $V_{CEsat} \approx 0.2$ V we could calculate the effective current gain from the expression given above:

$$0.2 = 20 - 500\beta \, (0.19 \text{ mA}) \left( 1 + \frac{\beta+1}{\beta} \right)$$

$$\Rightarrow \quad \beta = 104$$

(c) $I_C = \beta I_B = 6$ mA, $V_{CE} = 13.98$ V

**7.7**   The circuit is shown in Figure P8.7.

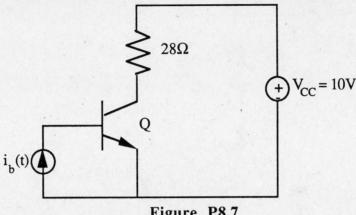

**Figure  P8.7**

Using the characteristic curves of a 2N3904 transistor, determine

      a)  The operating point of the transistor, Q, if $i_b(t) = 2.15$ mA $+ \Delta I_B$ and $\Delta I_B$    very small.

      b)  What is $\beta$ at this point?

      c)  Estimate $h_{fe}$ for this circuit at the operating point.  Also estimate $h_{ie}$.

**Solution:**

b)  Assume a typical $\beta$ ($\approx h_{FE}$) of 150 (from the data sheets of p. 433).

a)  The operating point of the transistor is obtained as follows:

$$I_{CQ} = \beta \, I_{BQ} = 3.15 \text{ mA}$$
$$V_{CEQ} = 10 - 28 \times I_{CQ} = 9.9118 \text{ V}$$

c)  $h_{fe} \approx 150$.  The value of $h_{ie}$ at this operating point is around 4.5 $\Omega$  (from the data sheet)

**7.8** For the circuit of Figure 8.11, $V_{BB} = V_{CC} = 5$ V, $R_C = 1$ k$\Omega$, $R_E = 1$ k$\Omega$, $\beta = 80$. Find the operating state of the transistor if

(a) $R_B = 100$ k$\Omega$

(b) $R_B = 10$ k$\Omega$

(c) $R_B = 1$ k$\Omega$

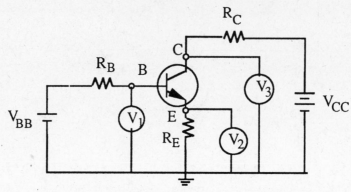

**Figure 8.11**

**Solution:**

By KVL, we have

$$5 = R_B I_B + V_{BE} + 1000(I_B + I_C)$$
$$5 = 1000 I_C + V_{CE} + 1000(I_B + I_C)$$

For operation in the active region, we assume

$$V_{BE} = 0.7 \text{ V}, \ I_C = \beta I_B, \ V_{CE} > 0.2 \text{ V}$$

For operation in the saturation region,

$$V_{BE} = 0.7 \text{ V}, \ V_{CE} = 0.2 \text{ V}, \ I_C < \beta I_B$$

(a) For the case $R_B = 100 \text{ k}\Omega$, assuming active region, we have

$$I_B = \frac{4.3}{100 \text{ k}\Omega + 80 \times 1 \text{ k}\Omega} = 23.9 \ \mu A$$

$$I_C = 80 I_B = 1.911 \text{ mA}$$

$$V_{CE} = 5 - 2000 I_C - 1000 I_B = 1.174 \text{ V}$$

Since $V_{CE} > 0.2$ V, the transistor is truly in the active region, and our results are valid.

(b) For $R_B = 10 \text{ k}\Omega$, assuming operation in the saturation region, we find:

$$5 - 0.7 = 11000 I_B + 1000 I_C$$
$$5 - 0.2 = 1000 I_B + 2000 I_C$$

and

$$I_B = 181 \ \mu A \quad I_C = 2.31 \text{ mA}$$

Since the actual $\beta = \frac{I_C}{I_B} = 12.76 < 80$, the transistor is in fact in the saturation region.

(c) For $R_B = 1 \text{ k}\Omega$, assuming operation in the saturation region, we have

$$4.3 = 2000 I_B + 1000 I_C$$
$$4.8 = 1000 I_B + 2000 I_C$$

$$I_B = 1.267 \text{ mA} \qquad I_C = 1.767 \text{ mA}$$

Here,

$$\beta = \frac{I_C}{I_B} = \frac{1.767}{1.267} = 1.39 < 80$$

so the assumption is correct. The transistor must be in saturation.

**7.9** Find the β for the each transistor shown in Figure P8.9.

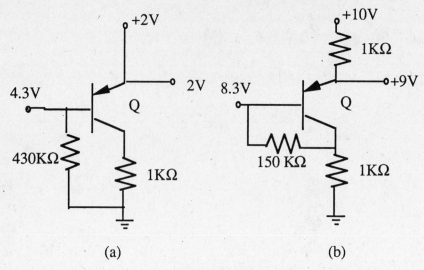

(a)                                        (b)

**Figure P8.9**

**Solution:**

(a) We have

$$V_B = 4.3 \text{ V}, \quad V_C = 2 \text{ V}$$

The collector current is

$$I_C = \frac{2}{1 \text{ k}\Omega} = 2 \text{ mA}$$

The base current $I_B$ is

$$I_B = \frac{4.3}{430 \text{ k}\Omega} = 0.01 \text{ mA}$$

Therefore, $\beta$ is approximately equal to:

$$\beta = \frac{I_C}{I_B} = 200$$

(b) The emitter current is

$$I_E = \frac{10 - 9}{1 \text{ k}\Omega} = 1 \text{ mA}$$

The voltage at the collector is

$$V_C = 8.3 \frac{1}{150 + 1} = 0.055 \text{ V}$$

The base current is

$$I_B = \frac{8.3 - 0.055}{150} = 55 \text{ }\mu\text{A}$$

Therefore, the collector current is

$$I_C = I_E - I_B = 1 - 0.055 = 0.945 \text{ mA}$$

and $\beta$ is approximately equal to:

$$\beta = \frac{I_C}{I_B} = \frac{0.945}{0.055} = 17.18$$

---

**7.10** For the circuit shown in Figure P8.11, $R_E = 200\Omega$, $R_1 = 11.25 \text{ k}\Omega$, $R_2 = 10 \text{ k}\Omega$.

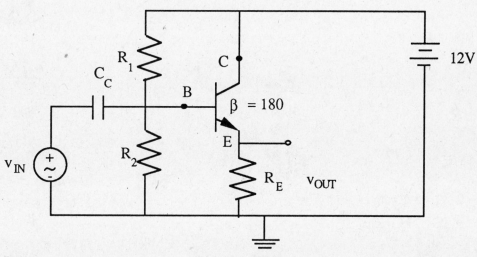

**Figure P8.11**

a) Find the operating point of the transistor.

b) Draw the h parameter circuit.

c) Determine the voltage gain $\dfrac{V_{OUT}}{V_{IN}}$ for $h_{fe} = \beta$, $h_{oe} = 10^{-4}$ $\Omega$

**Solution:**

(a) The voltage at the base is

$$V_B = \frac{R_2}{R_2 + R_1}\, 12 = 5.65 \text{ V}$$

Assuming $V_{BE} = 0.6$ V, we have

$$V_E = 5.05 \text{ V and } I_E = \frac{V_E}{200} = 25.25 \text{ mA}$$

Therefore, the base current is

$$I_B = \frac{I_E}{\beta + 1} = 0.14 \text{ mA}$$

The collector current is

$$I_C = \beta I_B = 25.2 \text{ mA}$$

Also

$$V_{CE} = V_C - V_E = 12 - 5.05 = 6.95 \text{ V}$$

(b) $h_{ie} = \frac{\partial V_{BE}}{\partial i_B}\Big|_{I_{BQ}}$

$\approx \frac{0.6}{0.14\times10^{-3}} = 4.29$ k$\Omega$

$$h_{fe} \approx \beta = 180$$

The h-parameter model is shown below

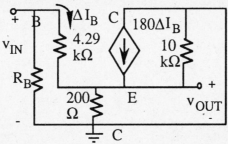

(c) We have R = 200 ‖ 10 k$\Omega$

$= \frac{10\times10^3\times200}{10\times10^3 + 200} = 196.08$ $\Omega$

$V_{in} = \Delta I_B 4.29\times10^3 + (\Delta I_B +$

$+ 180\Delta I_B)196.08$

$V_{out} = (\Delta I_B + 180\Delta I_B)196.08$

Therefore, AC the voltage gain is

$A_V = \frac{V_{out}}{V_{in}}$

$= \frac{181\times196.08}{181\times196.08 + 4290} = 0.89$

---

**7.11**  Design the bias circuit of the common-emitter amplifier in Figure 8.20 to establish an enitter current $I_E$ = 1mA.  Assume that the power supply $V_{CC}$ is 15 V,  and that $V_{BB} = \frac{1}{3}V_{CC}$,  and

$I_C R_C = \frac{1}{3} V_{CC}$.

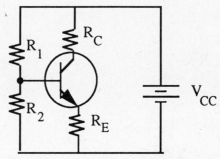

**Figure 8.20   Practical BJT self-bias DC circuit**

**Solution:**

For $V_{BB} = \dfrac{R_2}{R_1 + R_2} V_{CC} = \dfrac{1}{3} V_{CC}$

we must have

$$\frac{R_2}{R_1 + R_2} = \frac{1}{3}$$

We can select $R_2 = 10 \ k\Omega$, $R_1 = 20 \ k\Omega$.

Also

$$V_E = V_{CC} - V_{CE} - I_C R_C = \frac{1}{3} V_{CC} = 5 \ V$$

and $\qquad I_E = 1 \ mA$

Therefore,

$$R_E = \frac{V_E}{I_E} = 5 \ k\Omega$$

Assuming $I_C \approx I_E$, we have

$$I_C R_C = \frac{1}{3} V_{CC} = 5 \ V$$

$$R_C = \frac{5}{I_C} = 5 \ k\Omega$$

---

**7.12** Some switching circuits require a higher voltage or current than may be available from a computer or another source. A power transistor drive circuit may be used to provide a "buffer" between the computer and load as shown in Figure P8.17(a).

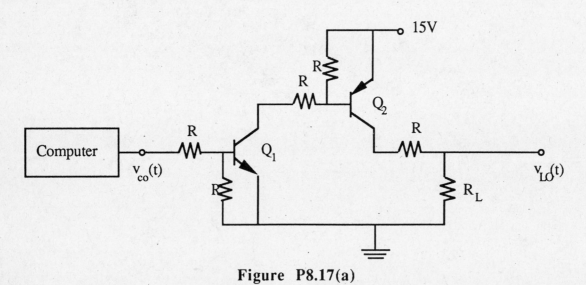

**Figure P8.17(a)**

If the waveform of the computer's output voltage, $V_{co}(t)$, is as shown in Figure P8.17(b), determine the output voltage $v_{LO}(t)$, given the following conditions.

$V_{CEsat} = \pm 0.2$ V     (+ for *npn*      - for *pnp*)

$V\gamma = \pm 0.8$ V   (+ for *npn*      - for *pnp*)

$R = 10\ \Omega$

$R_L = 50\ \Omega$

$B_{sat} = 30\ \Omega$

The waveform of $v_{co}(t)$ is shown in Figure P8.19.2.

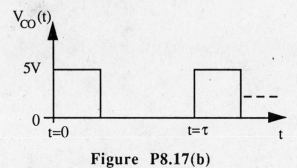

**Figure  P8.17(b)**

**Solution:**

When $V_{CO}$ is 0, transistors $Q_1$ and $Q_2$ are in the cut-off region. When $V_{CO}$ is applied to the circuit,

$$V_{BEQ1} = 0.8 \text{ V}$$

and

$$I_{BQ1} = \frac{V_{CO} - 0.8}{R} = 0.42 \text{ A}.$$

Thus, $Q_1$ will be in saturation. The voltage at the base of $Q_2$ is $(15 - 0.2)\dfrac{R}{R + R} = 7.4$ V. Assuming $Q_2$ is in the saturation region, the voltage at the collector would be 14.8 V. Thus, $V_{CBQ2} = V_{CQ2} - V_{BQ2} = 14.8 - 7.4 = 7.4$ V. Therefore, the CB junction and EB junction of $Q_2$ are forward biased and $Q_2$ is in the saturation region. The output voltage is:

$$V_{LO} = (15 - 0.2)\frac{R_L}{R + R_L} = 12.33 \text{ V}$$

The output waveform is shown below

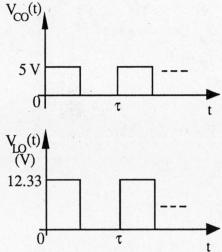

# Section 8: Field-Effect Transistors and Power Electronics

**8.1** Determine the operating state for a JFET having:

(a) $V_{DS} = -4$ V, $V_{GS} = 2$ V, $V_p = 1$ V

(b) $V_{DS} = -4$ V, $V_{GS} = 1$ V, $V_p = 2$ V

(c) $V_{DS} = 2$ V, $V_{GS} = 0$ V, $V_p = -3$ V

(d) $V_{DS} = 4$ V, $V_{GS} = -2$ V, $V_p = -4$ V

(e) $V_{DS} = 2$ V, $V_{GS} = -2$ V, $V_p = -1$ V

(f) $V_{DS} = -3$ V, $V_{GS} = 1$ V, $V_p = 4$ V

**Solution:**

(a)  In a p-channel JFET, $V_p$ is positive,

$0 \leq V_{GS} \leq V_p$, and $V_{DS}$ is negative.

To turn on the transistor, we require $V_{GS} < V_p$.  Since $V_{GS} = 2$ V $> V_p$, the JFET is in the cut-off region.

(b)  In a p-channel JFET, $V_p$ is positive, $0 \leq V_{GS} \leq V_p$ and $V_{DS}$ is negative.  In this case,

$$V_{DS} = -4 \text{ V} < V_{GS} - V_p = 1 - 2 = -1 \text{ V}$$

Therefore, the JFET is in the saturation region.

(c)  This is an n-channel JFET.

$V_{DS} = 2$ V, $V_{GS} = 0$ and $V_p = -3$ V

Therefore,

$$V_{DS} = 2 < 0 - (-3) = 3 \text{ V}$$

The JFET is in the ohmic region.

(d)  This is an n-channel JFET.

$V_{DS} = 4$ V, $V_{GS} = -2$ V and $V_p = -4$ V

Therefore,

$$V_{DS} = 4 \text{ V} > -2 - (-4) = 2 \text{ V}$$

The JFET is in the saturation region.

(e)  This is an n-channel JFET.

$V_{DS} = 2$ V, $V_{GS} = -2$ V and $V_p = -1$ V

We have

$$V_{GS} < -V_p$$

Therefore, the JFET is in the cut off region.

(f)  For a p-channel JFET, $V_p$ is positive,

$0 \leq V_{GS} \leq V_p$ and $V_{DS}$ is negative.

$$V_{DS} = -3 \text{ V} = V_{GS} - V_p = 1 - 4 = -3 \text{ V}$$

Therefore, the JFET is in either the ohmic or saturation region.

**8.2** Place the correct circuit symbol for a JFET in each circle in Figure P9.1, showing the proper connections to the external terminals. Identify the source and drain. Assume operation in the saturation region.

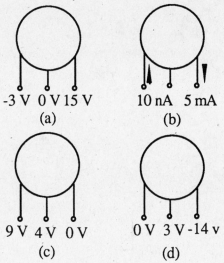

-3 V  0 V 15 V
(a)

10 nA   5 mA
(b)

9 V  4 V  0 V
(c)

0 V  3 V -14 v
(d)

**Figure  P9.1**

## Solution:

(a) The circuit is shown below.

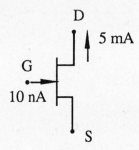

D
15 V

G
- 3 V

S
0 V

(b) The circuit is shown below

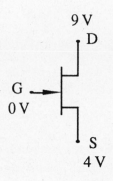

D
5 mA

G
10 nA

S

(c) The circuit is shown below

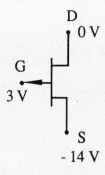

9 V
D

G
0 V

S
4 V

(d) The circuit is shown below

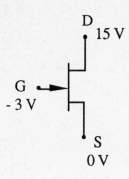

D
0 V

G
3 V

S
- 14 V

**8.3**    A n-channel enhancement MOSFET is measured to have a drain current of 4 mA at $V_{GS} = V_{DS} = 10$ V and of 1mA at $V_{GS} = V_{DS} = 6$ V.  Find parameters K and $V_T$.

**Solution:**

Assuming

$$v_{DS} > v_{GS} - V_T$$

we can calculate the constants applying equation 9.12 twice:

$$0.004 = K(10 - V_T)^2 \qquad (1)$$

$$0.001 = K(6 - V_T)^2 \qquad (2)$$

Equation (2) divided by equation (1) yields

$$4 = \frac{(10 - V_T)^2}{(6 - V_T)^2}$$

Solving the above equation, we have

$$V_T = 2 \text{ V}$$

From (1), $0.004 = K(10 - 2)^2$, we have

$$K = \frac{0.004}{64} = 62.5 \times 10^{-6} \frac{A}{V^2}$$

---

**8.4**  An enhancement type NMOS transistor has $V_T = 2$ V, $I_D = 1$ mA when $V_{GS} = V_{DS} = 3$ V. Neglecting the dependence of $I_D$ on $V_{DS}$ in saturation, find the value of $I_D$ for $V_{GS} = 4$ V.

**Solution:**

Because $v_{DS} > v_{GS} - V_T$, the transistor is in the saturation region.

$$i_D = K(v_{GS} - V_T)^2 = K(3 - 2)^2 = 0.001$$

$$K = 0.001 \frac{A}{V^2} = 1 \frac{mA}{V^2}$$

For $v_{GS} = 4$ V, we have

$$i_D = 0.001(4 - 2)^2 = 0.004 \text{ A} = 4 \text{ mA}$$

---

**8.5** An n-channel enhancement mode MOSFET is operated in the ohmic region, $V_{DS} = 0.4$ V, $V_T = 3.2$ V. The effective resistance of the channel is given by $R_{SD} = \dfrac{500}{V_{GS} - 3.2}$ ohms. Find $I_D$ when

(a) $V_{GS} = 5$ V

(b) $R_{SD} = 500\ \Omega$

(c) $V_{GD} = 4$ V

**Solution:**

Since

$$V_{DS} = 0.4\text{ V} < V_{GS} - V_T = 5 - 3.2 = 1.8\text{ V}$$

the transistor is operating in the triode region.

The effective resistance is

$$R_{SD} = \frac{500}{5 - 3.2} = 277.78\ \Omega$$

Since $R_{DS} = \dfrac{V_{DS}}{I_D}$, we have

$$I_D = \frac{V_{DS}}{R_{DS}} = \frac{0.4}{277.78} = 1.44\text{ mA}$$

---

**8.6** A JFET having $V_p = -2$ V and $I_{DSS} = 8$ mA is operating $V_{GS} = -1$ V and a very small $V_{DS}$. Find

a) $r_{DS}$

b) $V_{GS}$ at which $r_{DS}$ is half of its value in (a)

**Solution:**

(a) $r_{DS} = \dfrac{-V_D{}^2}{2I_{DSS}(V_p - V_{GS})}$

$= \dfrac{-(-2)^2}{2(0.008)(-2-(-1))} = 250 \ \Omega$

(b) $\dfrac{1}{2} \ 250 = \dfrac{-4}{2 \times 0.008 \times (-2 - V_{GS})}$

Therefore, $V_{GS} = 0 \ V$

---

**8.7** For the circuit shown in Figure 10.17, if the LR load is replaced by a capacitor, draw the output waveform and label the values.

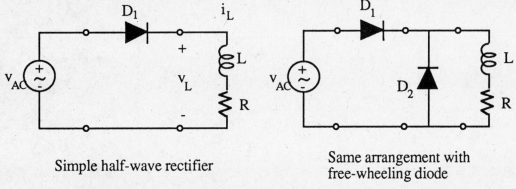

Simple half-wave rectifier          Same arrangement with
free-wheeling diode

**Figure 10.17 Rectifier connected to inductive load**

**Solution:**

When the sinusoidal source voltage is in the positive half cycle, the series diode conducts, and the shunt diode is an open circuit; thus, the positive half cycle appears directly across the capacitor (assuming ideal diodes). During the negative half cycle, the series diode is open, and therefore the voltage across the capacitor remains zero, as shown in the sketches below.

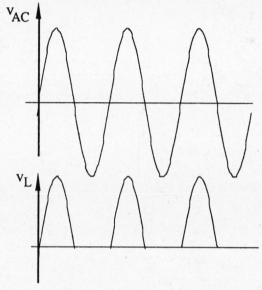

**8.8** Draw $v_L(t)$ and label the values for the circuit in Figure 10.17, if the diode forward resistance is 50 Ω, the forward bias voltage is 0.7 V, and the load consists of a resistor R = 10 Ω and an inductor L = 2H.

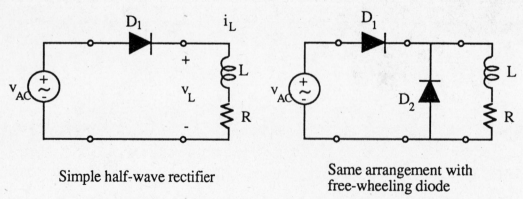

Simple half-wave rectifier

Same arrangement with free-wheeling diode

**Figure. 10.17 Rectifier connected to inductive load**

**Solution:**

To obtain exact numerical values, we assume a 110 Vrms source, R = 10 Ω, and L = 2 H; then:

$$v_{AC}(t) = A \sin (\omega t) = 155.6 \sin (377 t)$$

and from equation 9.42, the average load current is:

$$I_L = \frac{155.6}{\pi R} = 4.95 \text{ A}$$

Using the approximation

$$v_L(t) \approx \frac{A}{2} + \frac{A}{2} \sin \omega t$$

we have

$$v_L(t) \approx \frac{A}{2} + \frac{A}{2} \sin \omega t = 77.8 + 77.8 \sin(377 t)$$

The waveform is shown below.

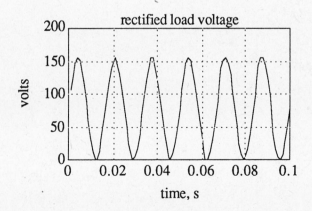

**8.9** For the circuit shown in Figure P9.11, $v_{AC}$ is a sinusoid with 10 V peak amplitude, $R = 2$ k$\Omega$, and the forward-conducting voltage of D is 0.7 V

    (a) sketch the waveform of $v_L(t)$

    (b) find the average value of $v_L(t)$

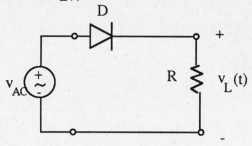

**Figure P9.11**

**Solution:**

(a) Assume $v_{AC} = 10 \sin\omega t$ V

The output voltage is

$$v_L(t) = (10 - 0.7) \sin\omega t$$

The waveform is shown below.

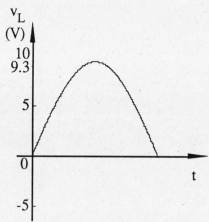

(b) $<v_L> \approx \dfrac{1}{2\pi} \displaystyle\int_0^\pi 9.3\sin(\omega t)\, d(\omega t)$

$\qquad = \dfrac{9.3}{\pi} = 2.96$ V

## Section 9: Operational Amplifiers

**9.1** Determine the output of the differential amplifier shown in Figure P10.2. Note that this amplifier is designed to amplify the difference between two signals.

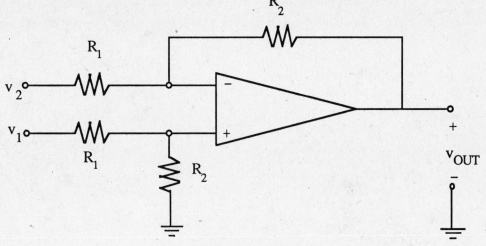

**Figure P10.2**

**Solution:**

Applying the voltage divider rule at the non-inverting terminal,

$$v^+ = \frac{R_2}{R_1 + R_2}v_1$$

Applying KCL at the inverting terminal,:

$$\frac{v^- - v_2}{R_1} + \frac{v^- - v_{out}}{R_2} = 0$$

Since $v^- = v^+$,     $v_{out} = \frac{R_2}{R_1}(v_1 - v_2)$

**9.2**   In the circuit of Figure P10.4, find the current i.

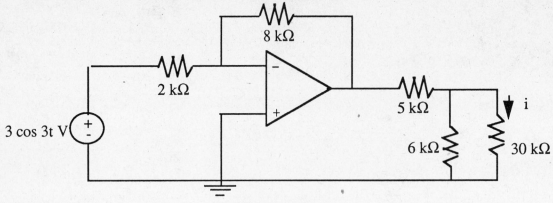

**Figure P10.4**

**Solution:**

We can find *i* using the following circuit:

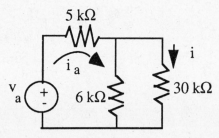

where $v_a$ is the output of the op-amp:

$$v_a = -\frac{8}{2}(3 \cos 3t) = -12 \cos 3t \text{ V}.$$

Using mesh current analysis,

$$11i_a - 6i = -12 \times 10^{-3} \cos 3t$$

$$36i - 6i_a = 0$$

where $i_a$ is the current in the first mesh. Solving the two equations,

$$i = -\frac{1}{5} \cos 3t \text{ mA}$$

**9.3** In the circuit of Figure P10.8, find the voltage $v$.

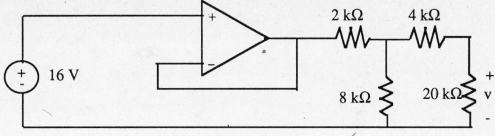

**Figure P10.8**

**Solution:**

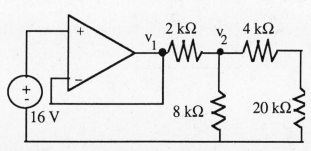

The first stage of the circuit is a voltage follower. Therefore,

$$v_1 = 16 \text{ V}$$

Using nodal analysis:

$$\left(\frac{1}{2000}+\frac{1}{8000}+\frac{1}{4000}\right)v_2 - \frac{1}{2000} v_1 - \frac{1}{4000} v = 0$$

$$\left(\frac{1}{4000}+\frac{1}{20000}\right)v - \frac{1}{4000} v_2 = 0$$

Solving for v, we find: $\qquad v = 10 \text{ V}$

**9.4** For the circuit of Figure P10.11, find the voltage v and the current i.

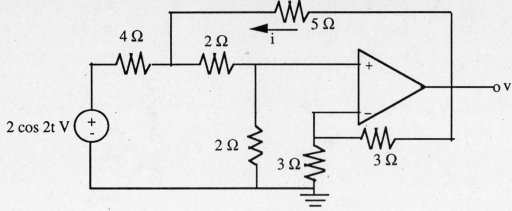

**Figure P10.11**

**Solution:**

The circuit is shown below:

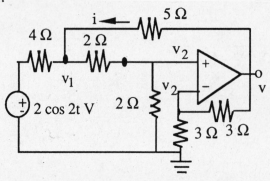

Applying nodal analysis,

$$(\frac{1}{4} + \frac{1}{5} + \frac{1}{2}) v_1 - \frac{1}{2}v_2 - \frac{1}{5} v = \frac{1}{2} \cos 2t$$

$$(\frac{1}{2} + \frac{1}{2})v_2 - \frac{1}{2}v_1 = 0$$

$$(\frac{1}{3} + \frac{1}{3})v_2 - \frac{1}{3} v = 0$$

$$v_1 = v = \cos 2t \text{ V}$$

and  $v_2 = 0.5 \cos 2t \text{ V}$

Therefore  $i = \frac{v - v_1}{5} = 0 \text{ A}$

**9.5** For the circuit of Figure P10.12, find the current $i$.

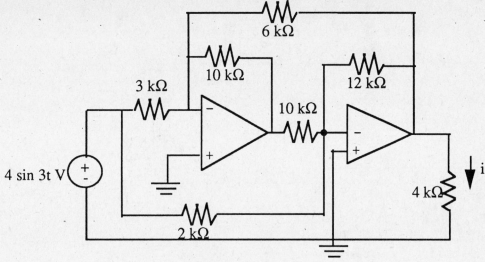

**Figure P10.12**

**Solution:**

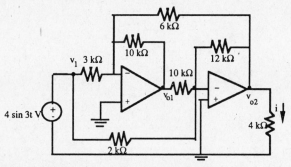

Applying KCL at the inverting terminals of the two op-amps,

$$-\frac{1}{3000} v_1 - \frac{1}{10000} v_{o1} - \frac{1}{6000} v_{o2} = 0$$

$$-\frac{1}{2000} v_1 - \frac{1}{10000} v_{o1} - \frac{1}{12000} v_{o2} = 0$$

$$v_1 = 4 \sin 3t$$

Solving for $v_{o1}$ and $v_{o2}$,

$$v_{o2} = -\frac{80}{3} \sin 3t , \; v_{o2} = 8 \sin 3t$$

$$i = \frac{v_{o1}}{4000} = 2 \sin 3t \; mA$$

**9.6** For the circuit of Figure P10.13, determine the equivalent resistance of the inverting amplifier $R_{EQ}$ seen by the source. [Hint: replace the signal source with an ideal 1A current source, and track the path of the current].

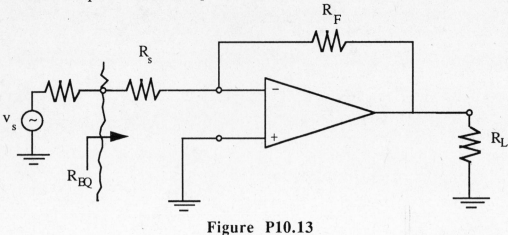

**Figure P10.13**

**Solution:**

With $v^+ = v^- = 0$ V

$$R_{eq} = \frac{v_i}{I_s}$$

$$I_s = \frac{v_i - v^-}{R_s} = \frac{v_i - v^+}{R_s} = \frac{v_i}{R_s}$$

Therefore,      $R_{eq} = R_s$

---

**9.7** Use superposition to determine an expression for the output voltage of the circuit of Figure P10.19.

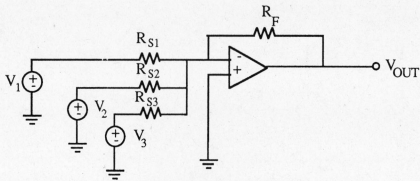

**Figure P10.19**

**Solution:**

The nature of the problem appears more clearly if we consider the following equivalent circuit, in which $R_{Sn}$ represents any one of the three source resistors, and $R_S'$ is the parallel combination of the other two, where the voltage sources have been replaced by short circuits.

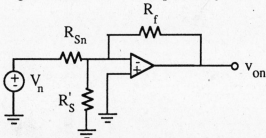

We note that there is no voltage drop across $R_S'$ since the inverting and noninverting nodes are at the same potential.

$$V_{on} = - \frac{R_f}{R_{Sn}} V_n$$

Thus, the output voltage due to all three sources is the weighted sum of the three.

$$V_o = V_{o1} + V_{o2} + V_{o3}$$

$$V_o = \frac{-R_f}{R_{S1}} V_1 - \frac{R_f}{R_{S2}} V_2 - \frac{R_f}{R_{S3}} V_3$$

**9.8** Assume that the op-amp in circuit of Figure P10.20 is ideal, that is, $A_V(OL) \to \infty$.

    a) Use superposition to find $v_{OUT}$ as a function of $v_2$.

    b) Again using superposition find $v_{OUT}$ due to $v_1$.

    c) If $v_2 = 5$ V and $v_1 = 4$ V, what is $v_{OUT}$?

    d) Determine a general expression for $v_{OUT}$.

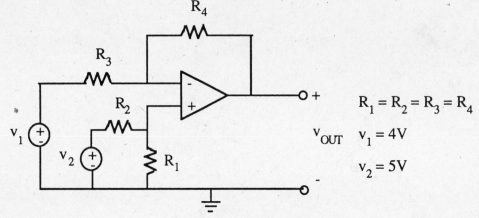

$R_1 = R_2 = R_3 = R_4$

$v_1 = 4V$

$v_2 = 5V$

**Figure P10.20**

**Solution:**

  a) $V_2$ sees a non inverting amplifier circuit:

$$A_{CL} = \left(\frac{R_3 + R_4}{R_3}\right)\frac{R_1}{R_1 + R_2}$$

$$= \frac{2R}{R}\frac{R}{2R} = 1$$

$$v_o|v_2 = v_2$$

b) $V_1$ sees an inverting amplifier circuit:

$$A_{CL} = -\frac{R_4}{R_3} = -1$$

$$v_o|v_1 = -v_1$$

c) $v_o = v_o|v_1 + v_o|v_2 = -4 + 5 = 1$ V

d) $v_o = v_2 - v_1$

**9.9** Find the gain of the amplifier circuit in Figure P10.21.

Assume $R_1 = R_2 = R_3 = 3$ k$\Omega$ and $R_4 = 1$ 00k$\Omega$.

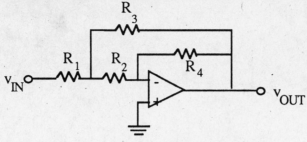

**Figure P10.21**

**Solution:**

Using the solution derived for Problem 10.23,

$$\frac{v_o}{v_i} = -\frac{R_{eq}}{R_1}\left(\frac{R_3R_4}{R_{eq}R_4 + R_2R_3}\right)$$

$$= \frac{1}{3}\left(\frac{(3)(100)}{(1)(100) + (3)(3)}\right)$$

$$= \frac{1}{3}\left(\frac{300}{100 + 9}\right) = \frac{100}{109}$$

$$\frac{v_o}{v_{in}} = 0.917$$

**9.10** Compute the frequency response of the circuit shown in Figure P10.26.

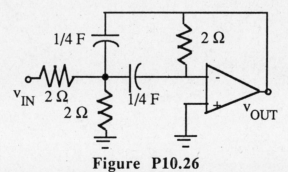

**Figure P10.26**

**Solution:**

Applying KCL at the indicated node and at the inverting terminal:

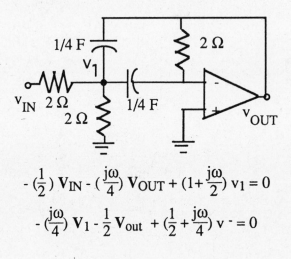

$$- \left(\frac{1}{2}\right) V_{IN} - \left(\frac{j\omega}{4}\right) V_{OUT} + \left(1 + \frac{j\omega}{2}\right) v_1 = 0$$

$$- \left(\frac{j\omega}{4}\right) V_1 - \frac{1}{2} V_{out} + \left(\frac{1}{2} + \frac{j\omega}{4}\right) v^- = 0$$

and since

$$v^- = 0 \text{ V}$$

Therefore,

$$\frac{V_{OUT}}{V_{IN}} = \frac{j2\omega}{\omega^2 - j4\omega - 8}$$

**9.11**  If a cosine wave of *2* V peak at *200* Hz is applied to the circuit of Figure P10.26, determine the output signal.

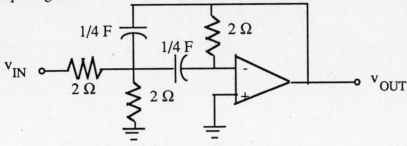

**Figure  P10.26**

**Solution:**

This problem is the same as example 10.8 except that

$$T = 1 \text{ s and } A = 2 \text{ V}$$

Therefore,

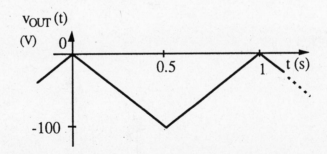

---

**9.12**  For an inverting amplifier with: $R_F = 47$ k$\Omega$, $R_S = 3.3$ k$\Omega$ and supply voltages $V_S^+ = 15$ V, $V_S^- = -12$ V.

a)  If the op-amp saturates at $0.9|V_S|$, plot the output voltage, $v_{OUT}(t)$, if:

$$v_S(t) = 0.5 + 0.5 \sin 2\pi t \text{ V}$$

b)  Repeat part a) if:

$$v_S(t) = -1 + 0.5 \sin 2\pi t \text{ Volts}$$

c) The slew rate of the op-amp is $2 \times 10^6$ V/sec. If:

$$v_S(t) = 0.5 \sin(4 \times 10^3 \pi t) \ Volts$$

Plot the output voltage $v_{OUT}(t)$ for the ideal case and for the slew rate-limited case.

**Solution:**

a) $v_0(t) = -\dfrac{47}{3.3} (0.5 + 0.5 \sin 2\pi t)$

$\qquad = -7.12 - 7.12 \sin 2\pi t$ (ideally)

op-amp saturates at 13.5 V and -10.8 V

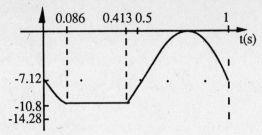

b) $v_o(t) = 14.24 - 7.12 \sin 2\pi t$ (ideally).

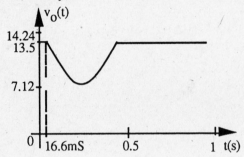

c) $v_o(t) = -7.12 \sin (4 \times 10^3 \pi t)$ V

$$S_{max} = 4 \times 10^3 \pi \times 7.12 = 8.95 \times 10^4 \text{ V/s}$$

The slew rate does not limit the output in this case. Therefore,

$$v_{o,actual} = v_{o,desired}$$

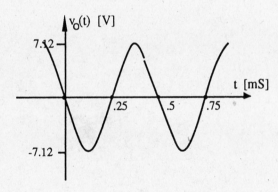

**9.13** The open loop gain, $A_{V(OL)}$, is actually a function of frequency. If we substitute the equation for the open loop frequency response into the closed loop gain equation for an inverting amplifier, we can obtain a more complete expression for the closed loop gain of the standard inverting amplifier, taking into account both the finite magnitude of $A_{V(OL)}$, as well as its limited frequency response.

a) Determine an expression of the form:

$$\frac{V_{OUT}}{V_{IN}}(j\omega) = \frac{K}{1 + \dfrac{j\omega}{\omega^*}}$$

where $K$ = gain

$\omega^*$ = new 3dB frequency

for the frequency response of the circuit shown, starting with the results of Problem 10.43. You may assume:

1) $R_F + R_S \ll A_o R_S$

or equivalently $\dfrac{R_F + R_S}{A_o R_S} \ll 1$

2) $\dfrac{R_F + R_S}{R_S A_o} \gg 1)$

b) Plot the frequency response $\left|\dfrac{V_{OUT}}{V_{IN}}\right|$ in dB vs frequency if

$$A_o = 10^5 \text{ and } f_o = 5 = \frac{\omega_o}{2\pi}.$$

Use the following cases and the equation you derived in part a)

| Case | $R_F$ | $R_S$ |
|------|-------|-------|
| i    | 10KΩ  | 10KΩ  |
| ii   | 100KΩ | 10KΩ  |
| iii  | 1MΩ   | 10KΩ  |
| iv   | 1MΩ   | 1KΩ   |

- plot all of these cases on the

  same set of axes with

  $1 \le f \le 1M$, $-10 \le |A_{CL}|_{dB} \le 100$ dB

- also plot the open loop gain $A_{V(OL)}(j\omega)$

  on these axes

9.14

**Solution:**

$$A_{CL} = -\frac{R_f}{R_s}\frac{1}{\left(1 + \dfrac{R_f + R_s}{R_s A_{OL}}\right)}$$

$$A_{OL} = \frac{A_o}{1 + j\dfrac{\omega}{\omega_o}}$$

Substituting the expression for $A_{OL}$,

$$\frac{1}{1 + \left(\dfrac{(R_f + R_s)/R_s A_o}{1 + j\dfrac{\omega}{\omega_o}}\right)} = \frac{1}{1 + \left(\dfrac{(R_f + R_s)(1 + j\dfrac{\omega}{\omega_o})}{R_s A_o}\right)}$$

$$= \frac{R_s A_o}{R_s A_o + R_f + R_s + j\dfrac{\omega}{\omega_o}(R_f + R_s)}$$

Now using the simplifications given in the problem statement, we obtain

$$A_{CL} \approx \left(-\frac{R_f}{R_s}\right)\left(\frac{R_s A_0}{R_s A_0 + j\frac{\omega}{\omega_0}R_f}\right)$$

$$= \left(-\frac{R_f}{R_s}\right)\left(\frac{1}{1 + j\frac{\omega}{\omega_0}\frac{R_f}{R_s A_0}}\right)$$

Therefore,

$$A_{CL} = \frac{v_0}{v_{in}} = -\frac{R_f}{R_s}\left(\frac{1}{1 + \frac{j\omega}{\omega_0 A_0 \frac{R_s}{R_f}}}\right)$$

and if we define

$$K = -\frac{R_f}{R_s}$$

$$\omega^* = \frac{\omega_0 A_0}{\frac{R_f}{R_s}}$$

we can write

$$A_{CL} = -\frac{R_f}{R_s}\left(\frac{1}{1 + \frac{j\omega}{\omega^*}}\right)$$

(b) The plots are given below.

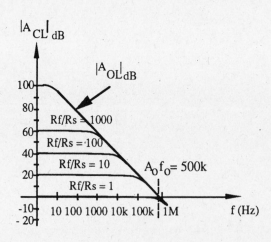

# Section 10: Integrated Circuit Electronics

**10.1** Derive the Common mode rejection ratio (CMRR) if the feedback resistance is $R_f + \Delta R$ and all other resistors are matched.

**Solution:**

The closed-loop gain is:

$$A = 1 + \frac{2R_2}{R_1}$$

The voltage at the non-inverting terminal is:

$$v^+ = \frac{R_F}{R_F + R} v_a{}'$$

The inverting-terminal voltage is:

$$v^- = v^+$$

The feedback current is given by:

$$i_f = \frac{v_{out} - v^-}{R_F + \Delta R} = \frac{v_{out} - \dfrac{R_F}{R_F + R} v_a{}'}{R_F + \Delta R}$$

The current going into the inverting terminal is:

$$i_s = \frac{v_b{}' - v^-}{R} = \frac{v_b{}' - \dfrac{R_F}{R_F + R} v_a{}'}{R}$$

Since $i_f = -i_s$,

$$\frac{v_b{}'}{R} - \frac{R_F}{R} \frac{1}{R_F + R} v_a{}' =$$

$$-\frac{v_{out}}{R_F + \Delta R} + \frac{R_F}{(R_F + R)(R_F + \Delta R)} v_a{}'$$

and

$$v_{out} = \frac{R_F}{R_F + R} v_a{}' + \frac{R_F}{R} \frac{R_F + \Delta R}{R_F + R} v_a{}' - \frac{R_F + \Delta R}{R} v_b{}'$$

$$= v_{out,dif} + v_{out,com}$$

Further,

$$v_a' = A(v_{a,dif} + v_{com}),$$
$$v_b' = A(v_{b,dif} + v_{com})$$

$$v_{out,com} = \frac{R_F}{R_F + R} A\, v_{com}$$

$$+ \frac{R_F}{R} \frac{R_F + \Delta R}{R_F + R} A\, v_{com} - \frac{R_F + \Delta R}{R} A\, v_{com}$$

$$= \frac{-\Delta R}{R_F + R} A\, v_{com}$$

The common-mode rejection ratio, CMRR,

is therefore given by:

$$CMRR_{dB} = 20\log_{10} \left| \frac{A_{dif}}{v_{outcom}/v_{com}} \right|$$
$$= 20\log_{10} \left| \frac{R_F + R}{\Delta R} \right|$$

---

**10.2** Derive the Common mode rejection ratio (CMRR) if the resistors $R_2$ and $R_2'$ are mismatched by $\Delta R$.

**Solution:**

$$A = 1 + \frac{2R_2 + \Delta R}{R},$$

$$v^+ = \frac{R_F}{R_F + R} v_a'$$

$$i_f = \frac{v_{out} - v^-}{R_F} = \frac{v_{out} - \frac{R_F}{R_F + R} v_a'}{R_F}$$

$$i_s = \frac{v_b' - v^-}{R} = \frac{v_b' - \frac{R_F}{R_F + R} v_a'}{R_F}$$

Since $i_f = -i_s$,

---

**10.3**  Derive the expression for the CMRR given by

$$CMRR_{dB} = \left| \frac{A_{dif}}{A_{com}} \right| = 20 \, log_{10} \left| \frac{A_{dif}}{\frac{V_{OUTcom}}{V_{com}}} \right|$$

$$= 20 log_{10} \left| \frac{R_F' + R'}{\frac{R_F}{R}\left(\frac{R_F'R}{R_F} - R'\right)} \right|$$

**Solution:**

The closed-loop gain is:

$$A = 1 + \frac{2R_2}{R_1}$$

$$v^+ = \frac{R_F'}{R_F' + R'}, \quad v_a' = v^-$$

$$i_f = \frac{v_{out} - v^-}{R_F} = \frac{v_{out}}{R_F} - \frac{R_F'}{R_F}\frac{v_a'}{R_F' + R'}$$

$$i_s = \frac{v_b' - v^-}{R} = \frac{v_b'}{R} - \frac{R_F'}{R}\frac{v_a'}{R_F'R'}$$

Since $i_f = -i_s$, we have

$$v_{out} = R_F\left\{\frac{R_F'}{R_F(R_F' + R')}v_a' - \frac{v_b'}{R} + \frac{R_F'}{R}\frac{v_a'}{R_F'R'}\right\}$$

where

$$v_a' = A(v_{a,dif} + v_{com}).$$
$$v_b' = A(v_{b,dif} + v_{com})$$

and

$$v_{out,com} = \frac{R_F'}{R_F'+R'} A \, v_{com}$$

$$-\frac{R_F}{R} A \, v_{com} + \frac{R_F'R_F}{R(R_F' + R')} A \, v_{com}$$

Therefore, the CMRR is:

$$CMRR = 20\log_{10}\left|\frac{A_{dif}}{v_{out,com}/v_{com}}\right| = 20\log_{10}\left|\frac{R_F' + R'}{\frac{R_F}{R}\left(\frac{R_F'R}{R_F} - R'\right)}\right|$$

**10.4** If $R_2 = 20$ kΩ for the IA of Figure 11.14, find $R_1$ so that the input stage gain of 300 can be achieved.

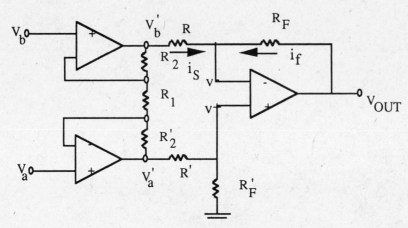

**Figure 11.14 Discrete op-amp instrumentation amplifier**

**Solution:**

$A = 1 + \dfrac{2R_2}{R_1}$, therefore, $300 = 1 + \dfrac{40}{R_1}$

Thus, $R_1 = 133.8 \; \Omega$.

---

**10.5** Find the differential gain for the IA of Figure 11.14 if $R_1 = 2 \; k\Omega$, $R_2 = R' = R = 10 \; k\Omega$, and $R_f = R'_f = 50 \; k\Omega$.

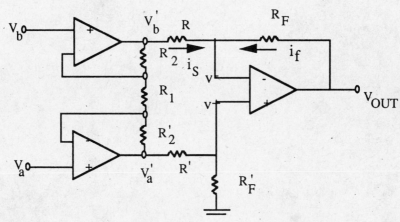

**Figure 11.14 Discrete op-amp instrumentation amplifier**

**Solution:**

$$A_{diff} = \frac{R_F}{R} \left( 1 + \frac{2R_2}{R_1} \right) = 55$$

---

**10.6** Let $K=1$ in a Sallen and Key low-pass filter prototype; if the cutoff frequency, $\omega_C$, and the resistors $R'=R_1=R_2$ and $R$ are normalized to unity, derive relationships for $C_1$, $C_2$, and $Q$.

**Solution:**

$$\frac{1}{Q} = \sqrt{\frac{R_2 C_2}{R_1 C_1}} + \sqrt{\frac{R_1 C_2}{R_2 C_1}} + (1\text{-}K)\sqrt{\frac{R_1 C_1}{R_2 C_2}}$$

Substituting the given values and simplifying the equation,

$$\frac{1}{Q} = 2\sqrt{\frac{C_2}{C_1}} \quad \text{or} \quad Q = \frac{1}{2}\sqrt{\frac{C_1}{C_2}}$$

---

10.5

**10.7**   Design a second-order Butterworth low-pass filter with *10* kHz cutoff frequency, a DC gain of *10*, $Q = 15$, and $V_S = \pm 15$ V.

**Solution:**

$$K = 1 + \frac{R_A}{R_B} = 10; \ R_A = 9 \ k\Omega \Rightarrow R_B = 1k\Omega$$

$$f = \frac{1}{2\pi\sqrt{R_1C_1R_2C_2}} = 10,000 \text{ Hz}$$

$$\left[\sqrt{\frac{R_2C_2}{R_1C_1}} + \sqrt{\frac{R_1C_2}{R_2C_1}} - 9\sqrt{\frac{R_1C_1}{R_2C_2}}\right] = 1/15$$

Choose $C_1 = C_2 = 1 \ \mu F$ and solve for $R_1 = 5.6 \ \Omega$ and $R_2 = 45.5 \ \Omega$. Then, substitute the values thus obtained in the low-pass filter.

---

**10.8**   Design a band-pass filter with the low cutoff frequency of *500* Hz, the high cutoff frequency of *12kHZ*, and a passband gain of *5*. Calculate the value of the $Q$ for the filter. Also, draw the approximate frequency response of this filter.

**Solution:**

$$Q = \frac{\sqrt{f_H f_L}}{f_H - f_L} = 2/3$$

$$K = 1 + \frac{R_A}{R_B} = 2;$$

$$\text{choose } R_A = 1 \ k\Omega \Rightarrow R_B = 1k\Omega$$

$$f = \frac{1}{2\pi\sqrt{R_1C_1R_2C_2}} = 2,000 \text{ Hz}$$

Choose R1 = R2 and C1 = C2 = 1 $\mu F$

$\Rightarrow R_1 = R2 = 80 \ \Omega$. Then, substitute the obtained values in the high-pass filter .

$K = 1 + \dfrac{R_A}{R_B} = 2.5$; choose $R_A = 1.5\ k\Omega$

$\Rightarrow R_B = 1k\Omega$

$f = \dfrac{1}{2\pi\sqrt{R_1C_1R_2C_2}} = 500\ Hz$

Choose $R_1 = R_2$ and $C_1 = C_2 = 1\ \mu F \Rightarrow$

$R_1 = R_2 = 318\ \Omega$. Then, substitute the obtained values in the low-pass filter of Figure. By connecting the output of the high-pass filter to the input of the low-pass filter, we obtain the desired filter.

---

**10.9**   Design two normalized low-pass quadratic filter sections with $Q$'s equal to *2* and *0.5*, respectively.  Plot the magnitude frequency response of each, making sure to plot enough points around the cutoff frequency.  Explain what is meant by "peaking" in the frequency response.

**Solution:**

$$K = 1$$

$$f = \frac{1}{2\pi\sqrt{R_1 C_1 R_2 C_2}} = \left( \sqrt{\frac{R_2 C_2}{R_1 C_1}} + \sqrt{\frac{R_1 C_2}{R_2 C_1}} \right) = 2$$

Choose $C_1 = C_2 = 1\ \mu F$, solve for $R_1$ and $R_2$. $R1 = R2 = 1\ k\Omega$. Thus, $\omega_c = 1{,}000$ rad/s.

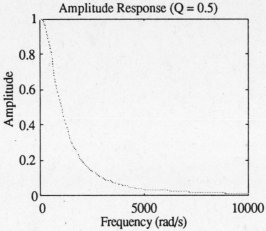

Amplitude Response (Q = 0.5)

$$K = 1,\ f = \frac{1}{2\pi\sqrt{R_1 C_1 R_2 C_2}} =$$

$$\left[ \sqrt{\frac{R_2 C_2}{R_1 C_1}} + \sqrt{\frac{R_1 C_2}{R_2 C_1}} \right] = 1/2 = 0.5$$

Choose $R_1 = R_2 = 1\ k\Omega$ and $C_1 = 16\ C_2$ where $C_1 = 1\ \mu F$. Therefore, $\omega_c = 250$ rad/s.

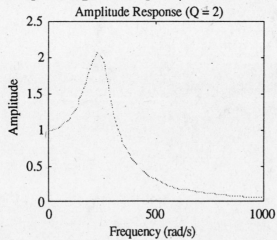

Amplitude Response (Q = 2)

"Peaking" refers to the resonant peak the filter frequency response displays for large values of Q.

**10.10**   In the circuit of Figure P11.3, $R_1 = 1$ $k\Omega$, $R_2 = 10$ $k\Omega$, $R_{om} = R_1 \| R_2$, $V_{in} = 5$ V peak-to-peak sine wave. Assuming supply voltages are $\pm 15$ V, determine the threshold voltages ( positive and negative $v^+$) and draw the output waveform.

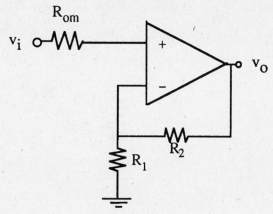

**Figure  P11.3**

**Solution:**

Applying KCL:

$$\frac{v_o}{v_{in}} = 1 + \frac{R_2}{R_1}$$

Therefore, $\frac{v_o}{v_{in}} = 11$

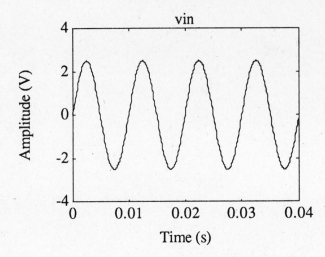

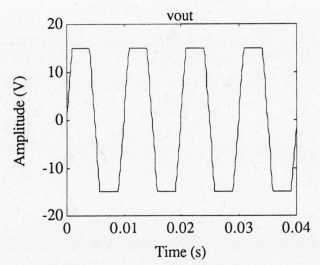

**10.11**  In the circuit of Figure P11.3, $R_1 = 2\ k\Omega$, $R_2 = 15\ k\Omega$, $R_{om} = R_1\|R_2$, $V_{in} = 2$V peak-to-peak sine wave. Assuming supply voltages are $\pm 15$ V, determine the threshold voltages ( positive and negative $v^+$) and draw the output waveform.

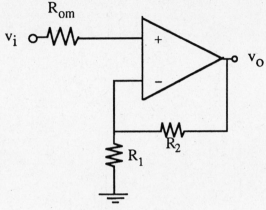

**Figure  P11.3**

**Solution:**

Applying KCL:

$$\frac{v_0}{v_{in}} = 1 + \frac{R_2}{R_1}$$

Therefore, $\frac{v_0}{v_{in}} = 8.5$

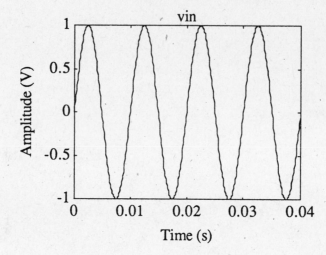

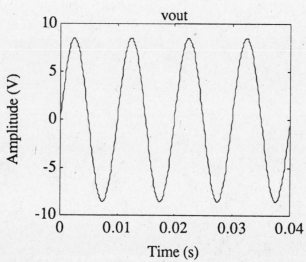

**10.12** For the circuit of Figure P11.5, $V_{in} = 100$ mV peak sine wave at 5 kHz, $R = 10$ k$\Omega$, $D_1$ is a 6.2 V Zener and $D_2$ is 0.7 V. Draw the output voltage waveform.

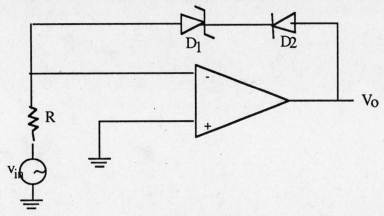

**Figure P11.5**

**Solution:**

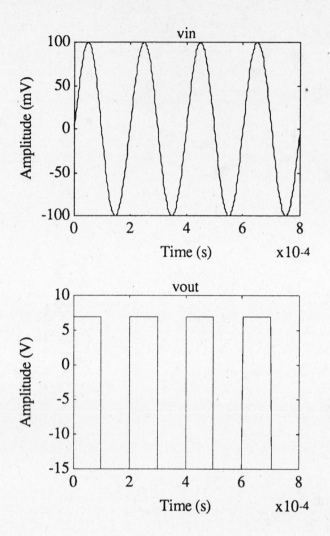

---

**10.13**  For the circuit of Figure P11.6, $V_{in} = 100$ mV peak sine wave at $5$ kHz, $R = 10$ k$\Omega$, $D_1$ and $D_2$ are $6.2$ V zeners.  Draw the output voltage waveform.

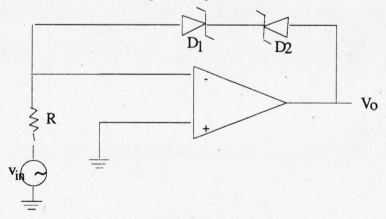

**Figure  P11.6**

10.14

**Solution:**

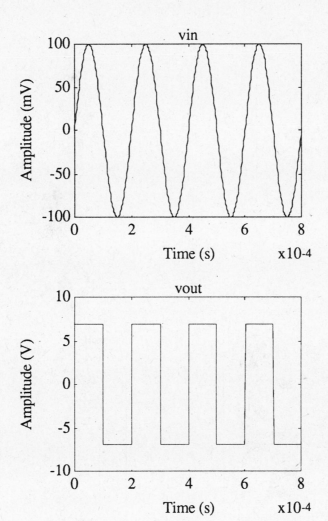

**10.14**   In Figure P11.7, the minimum value of $V_{IN}$ for a HIGH input is $2.7V$. Assume that transistor $Q_1$ has a $B$ of at least $50$. Find the range for resistor $R_B$ that can guarantee the transistor $Q_1$ to be ON.

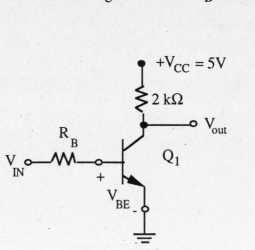

**Figure  P11.7**

**Solution:**

$i_C = \dfrac{5 - 0.2}{2000} = 2.4$ mA, therefore, $i_B = i_C/\beta = 48$ $\mu$A.  $(v_{in})_{min} = 2.7$ V and $(v_{in})_{max} = 5.0$ V,

therefore, applying KVL:

$$-v_{in} + R_B i_B + 0.6 = 0$$

or

$R_B = \dfrac{v_{in} - 0.6}{i_B}$, substituting for $(v_{in})_{min}$ and $(v_{in})_{max}$ , we find the following range for $R_B$:

$$43{,}750 \ \Omega \le R_B \le 91{,}667 \ \Omega$$

**10.15**  Figure P11.8 shows a very simple circuit with two transistor inverters connected in series where $R_{1C} = R_{2C} = 5k\Omega$ and $R_{1B} = R_{2B} = 12\ k\Omega$.

    a) Find $V_B$, $V_{OUT}$, and the state of transistor $Q_1$ when $V_{IN}$ is LOW.

    b) Find $V_B$, $V_{OUT}$, and the state of transistor $Q_1$ when $V_{IN}$ is HIGH.

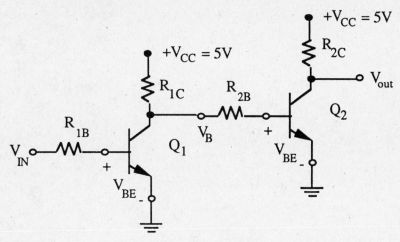

**Figure  P11.8**

**Solution:**

a) $v_{in}$ is low $\Rightarrow Q_1$ is cut off $\Rightarrow v_B = 5$ V $\Rightarrow Q_2$ is in saturation $\Rightarrow v_{out}$ = low = 0.2 V.

b) $v_{in}$ is high $\Rightarrow Q_1$ is in saturation $\Rightarrow v_B = 0.2$V $\Rightarrow Q_2$ is cut off $\Rightarrow v_{out}$ = high = 5 V.

**10.16**  For the inverter of Figure P11.9, $R_{C1} = R_{C2} = 2$ k$\Omega$ and $B_1 = B_2 = 10$. Show that transistor $Q_2$ saturates, and find a condition for $R_B$ so that $Q_1$ also saturates.

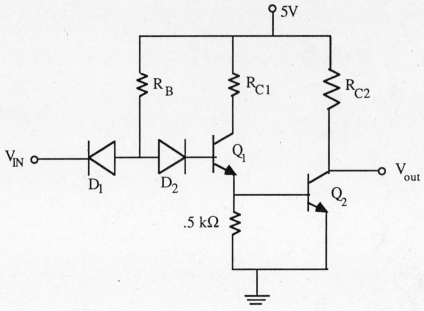

**Figure  P11.9**

**Solution:**

$i_{C2} = \dfrac{5 - 0.2}{2000} = 2.4$ mA, therefore, $i_{B2} = 2.4/\beta = 0.24$ mA. Applying KCL:

$i_{E1} = 0.24 + 0.12 = 0.36$ mA, therefore, $i_{B1} = 0.36/\beta = 0.036$ mA. Applying KVL:

$-5 + (.036)R_B + 1.8 = 0 \Rightarrow R_B = 89$ k$\Omega$

---

**10.17**   Figure P11.11 is a circuit diagram for a 3-input TTL NAND gate. Assuming that $V_1$ and $V_2$ are HIGH and $V_3$ is LOW, find $V_{B1}$, $V_{B2}$, $V_{B3}$, $V_{C2}$, and $V_{OUT}$. Also, indicate the state of each transistor.

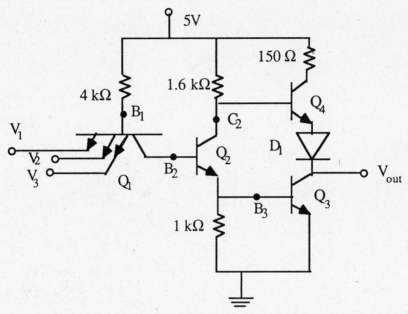

**Figure  P11.11**

**Solution:**

$Q_2$ and $Q_3$ are cut off and $Q_4$ conducts.

$v_{B1} = 0.6$ V, $v_{C2} = 5$ V, $v_{out} = 5$ V, $v_{B2} = 0$ V, $v_{B3} = 0$ V. The base of $Q_4$ goes to the supply potential which turns $Q_4$ on.

---

**10.18** For the circuit of Figure P11.11, discuss the purpose of the transistor $Q_4$ and Diode $D_1$.

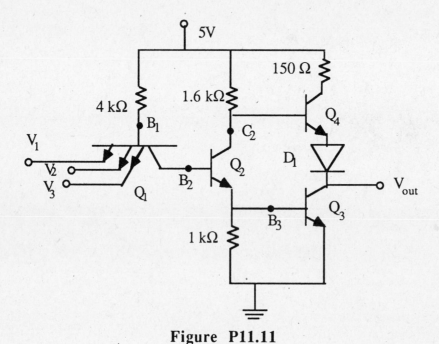

**Figure P11.11**

**Solution:**

$Q_4$ together with the 150 -k$\Omega$ resistor serves as a load for the output transistor $Q_3$. This yields a faster switching time. $D_1$ is used to ensure that $Q_4$ is off when $Q_2$ and $Q_3$ are off.

---

# Section 11: Digital Logic Circuits

**11.1** Using the method of proof by perfect induction show the validity of rule

$$(X+Y)(X+Z) = X + YZ$$

**Solution:**

Using proof by perfect induction ,

| X | Y | Z | (X+Y)(X+Z) | X + YZ |
|---|---|---|------------|--------|
| 0 | 0 | 0 | 0 | 0 |
| 0 | 0 | 1 | 0 | 0 |
| 0 | 1 | 0 | 0 | 0 |
| 0 | 1 | 1 | 1 | 1 |
| 1 | 0 | 0 | 1 | 1 |
| 1 | 0 | 1 | 1 | 1 |
| 1 | 1 | 0 | 1 | 1 |
| 1 | 1 | 1 | 1 | 1 |

we can see that the two expressions are equal.

**11.2** Using the method of proof by perfect induction, show the validity of rule

$$X + \bar{X}Y = X + Y$$

**Solution:**

Using proof by perfect induction,

| X | Y | $\bar{X}$ | X + Y | $X + \bar{X}Y$ |
|---|---|-----------|-------|----------------|
| 0 | 0 | 1 | 0 | 0 |
| 0 | 1 | 1 | 1 | 1 |
| 1 | 0 | 0 | 1 | 1 |
| 1 | 1 | 0 | 1 | 1 |

we can see that they are equal.

**11.3**   Using Demorgan's theorems and rules of Boolean algebra, simplify the following logic function.

$$F(X,Y,Z) = \overline{(X+Y)Z} + \overline{X\,\overline{\overline{YZ}}}$$

**Solution:**

$$F(X,Y,Z) = \overline{(X+Y)Z} + \overline{X\,\overline{\overline{YZ}}}$$

Applying De Morgan's theorems,

$$F = \overline{X+Y} + \bar{Z} + \overline{\overline{\overline{YZ}}} + \bar{X}$$

Applying De Morgan's theorems once more, and rearranging the terms

$$F = \bar{X} + \bar{X}\bar{Y} + YZ + \bar{Z}$$

Applying the rules of Boolean algebra,

$$F(X,Y,Z) = \bar{X} + Y + \bar{Z}$$

---

**11.4**   What is the logic function of the circuit of Figure P12.2?

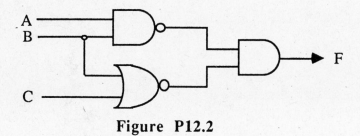

**Figure  P12.2**

**Solution:**

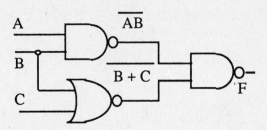

$$F = \overline{\overline{(B + C)} \; \overline{AB}} = B + C$$

---

**11.5** Using Karnaugh map, find the minimum expression for the output of the logic circuit shown in Figure P12.6.

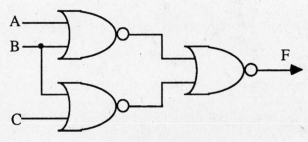

**Figure P12.6**

**Solution:**

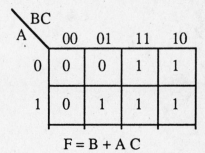

$$F = B + A\,C$$

---

**11.6** a) Fill in the Karnaugh Map for the logic function defined by the truth table of Figure P12.9.

| A | B | C | D | F |
|---|---|---|---|---|
| 0 | 0 | 0 | 0 | 1 |
| 0 | 0 | 0 | 1 | 1 |
| 0 | 0 | 1 | 0 | 1 |
| 0 | 0 | 1 | 1 | 1 |
| 0 | 1 | 0 | 0 | 0 |
| 0 | 1 | 0 | 1 | 1 |
| 0 | 1 | 1 | 0 | 0 |
| 0 | 1 | 1 | 1 | 1 |
| 1 | 0 | 0 | 0 | 1 |
| 1 | 0 | 0 | 1 | 1 |
| 1 | 0 | 1 | 0 | 0 |
| 1 | 0 | 1 | 1 | 0 |
| 1 | 1 | 0 | 0 | 0 |
| 1 | 1 | 0 | 1 | 1 |
| 1 | 1 | 1 | 0 | 1 |
| 1 | 1 | 1 | 1 | 0 |

**Figure  P12.9**

b) What is the minimum expression for the function?

c) Realize the function using only NAND gates.

**Solution:**

a)

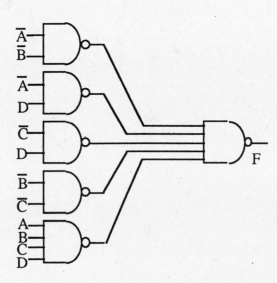

|     | 00 | 01 | 11 | 10 |
|-----|----|----|----|----|
| 00  | 1  | 1  | 1  | 1  |
| 01  | 0  | 1  | 1  | 0  |
| 11  | 0  | 1  | 0  | 1  |
| 10  | 1  | 1  | 0  | 0  |

b)  $\overline{A}\overline{B} + \overline{A}D + \overline{C}D + \overline{B}\overline{C} + ABC\overline{D}$

c)

**11.7**  Determine the minimum expression for the logic function

$$f(A,B,C,D) = A\overline{B} + AB\overline{C} + \overline{A}CD + AC\overline{D} + ABCD$$

**Solution:**

$$f(A,B,C,D) = A(\overline{B} + B\overline{C} + C\overline{D} + BCD) + \overline{A}CD$$

$$\underbrace{\qquad}_{1}$$

$$= A(1 + \overline{B} + D) + \overline{A}CD$$

$$\underbrace{\qquad}_{1}$$

$$= A + \overline{A}CD$$

$$= A + CD$$

---

**11.8**  What is the minimum expression for the output of the logic circuit shown in Figure P12.10?

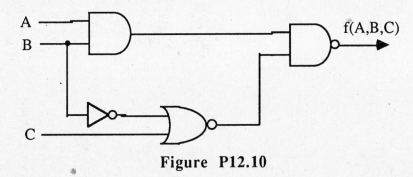

**Figure  P12.10**

**Solution:**

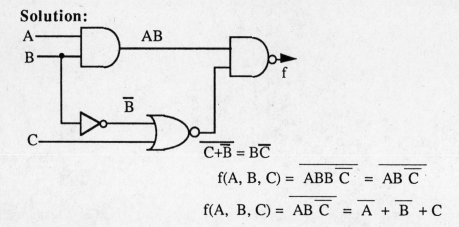

$$\overline{C+\overline{B}} = B\overline{C}$$

$$f(A, B, C) = \overline{ABB\,\overline{C}} = \overline{AB\,\overline{\overline{C}}}$$

$$f(A, B, C) = \overline{AB\,\overline{C}} = \overline{A} + \overline{B} + C$$

---

**11.9** Simplify the following expression:

$$(C + \overline{A}B)\left[(\overline{A}\overline{B} + B)C + A\right]$$

**Solution:**

$$F(A,B,C) = (C + \overline{A}B)\left[(\overline{A}\overline{B} + B)C + A\right]$$

$$= (C + \overline{A}B)(\overline{A}C + A) = (C + \overline{A}B)(C + A)$$

$$= C + \overline{A}BC + CA + \overline{A}BA$$

$$= C(\overline{A}B + A + 1)$$

$$= C$$

**11.10** A logic circuit is to have three inputs A, B, and C and one output F. The function F is to be a logic 1 if and only if an odd number of the input variables are at the logic 1 level. Make a truth table and Karnaugh Map for this function and find the minimum expression for F.

**Solution:**

| A | B | C | F |
|---|---|---|---|
| 0 | 0 | 0 | 0 |
| 0 | 0 | 1 | 1 |
| 0 | 1 | 0 | 1 |
| 0 | 1 | 1 | 0 |
| 1 | 0 | 0 | 1 |
| 1 | 0 | 1 | 0 |
| 1 | 1 | 0 | 0 |
| 1 | 1 | 1 | 1 |

| A \ BC | 00 | 01 | 11 | 10 |
|---|---|---|---|---|
| 0 | 0 | 1 | 0 | 1 |
| 1 | 1 | 0 | 1 | 0 |

$$F = \overline{A}\,\overline{B}\,C + \overline{A}\,B\,\overline{C} + ABC + A\,\overline{B}\,\overline{C}$$

The expression cannot be reduced further.

**11.11** Simplify the function (use Karnaugh Map) and draw a circuit.

$$F = \overline{A}C + A\overline{B}\overline{D} + \overline{A}C\overline{D} + \overline{A}B\overline{C}$$

**Solution:**

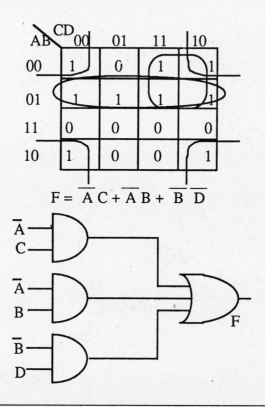

$$F = \overline{A}\,C + \overline{A}\,B + \overline{B}\,\overline{D}$$

11.10

**11.12**  a)  Complete the truth table for the circuit of Figure P12.16.

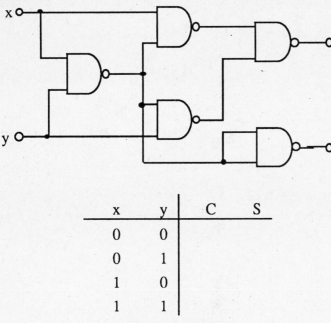

| x | y | C | S |
|---|---|---|---|
| 0 | 0 | | |
| 0 | 1 | | |
| 1 | 0 | | |
| 1 | 1 | | |

**Figure P12.16**

b)  What mathematical function does this circuit perform (Hint - binary operations) and what do the outputs signify?

c)  How many standard 14pin IC's would it take to construct this circuit?

**Solution:** a)

| x | y | C | S |
|---|---|---|---|
| 0 | 0 | 0 | 0 |
| 0 | 1 | 0 | 1 |
| 1 | 0 | 0 | 1 |
| 1 | 1 | 1 | 0 |

b)  A simple adder circuit.  The S output is the sum, while the C output contains the carry.

c)  2

**11.13** Find the simplified sum-of-products presentation of the function from the Karnaugh map shown in Figure P12.22.

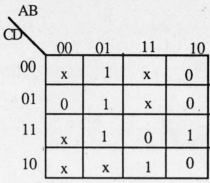

**Figure P12.22**

**Solution:**

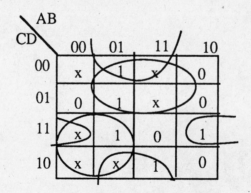

$$F = \bar{A}C + B\bar{C} + B\bar{D} + \bar{B}CD$$

**11.14**   A 256x4 memory can be organized using linear selection. Draw a block diagram of the logical organization of the memory.

**Solution:**

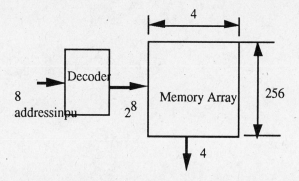

**11.15**   Draw a block diagram of internal structure of a 1,024x1 SRAM using linear selection.

**Solution:**

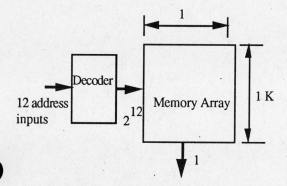

## Section 12: Digital Systems

**12.1** Draw a block diagram that shows the bus structure of a microprocessor based system and the subsystems connected by the bus connects. For each of subsystem indicate the nature of the components that make up the system, that is, ROM, RAM, etc.

**Solution:**

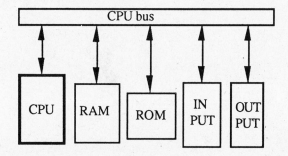

**12.2** Explain the purpose of the ALU.

**Solution:**
The arithmetic and logic unit (ALU) is where all computations take place, and is capable of performing the following operations on binary data:
• Binary addition and subtraction
• Logical AND, OR, EX-OR
• Complement
• Rotate left or right

**12.3**  Name the internal registers of a microprocessor and explain their functions.

**Solution:**

The internal registers include a program counter, the instruction register, general purpose registers and temporary registers. The sequence of instructions (often called micro instructions) is generated by a programming language, which is in principle similar to the FORTRAN, BASIC, or PASCAL languages. To keep track of which instruction is to be executed next, the control unit has a dedicated register, called the program counter (PC). The program counter holds the address of either the next instruction to be executed or the address of a multi-word instruction. When the control unit

requests the memory to transfer the data to the microprocessor, the data will be transferred into the microprocessor through the data bus latch and then into the instruction register. Registers inside the microprocessor are interconnected by an internal data bus. Temporary registers are used by the control unit to hold information until it is transferred to another register or used in a computation. To allow greater flexibility, instructions can effect transfers between general purpose registers. The contents of the registers are stored in a stack during subroutine calls and interrupts. The stack consists of a set of specifically allocated read/write memory; a stack pointer is required to address a location or register in the stack.

**12.4** Name the three different system buses and explain their functions.

**Solution:**

The system bus is divided into three busses each with a dedicated task. The Address Bus, which is a unidirectional bus, holds the address of the device with which the CPU wishes to communicate. The Data Bus, which is a bi-directional bus, transfers information to or from the device. The Control Bus contains the information about which operation is to be performed. This bus can be considered to act on the instruction, whereas the Address Bus controls the location of this action, and the Data Bus contains the result of the action or the input to the action.

---

**12.5** Explain function of the status register (flag register) and give an example.

## Solution:

The status register ( flag register) is the one used when acting on comparison statements, such as "If Greater Than ...", etc. The Status Register is a 16-bit register, however only 9 bits are utilized. The register looks as follows:

       x x x x O D I T S Z x A x P x C

The 'x' represents a bit which is not used, while the letters are bits which hold the following conditions:

| | |
|---|---|
| O | Overflow Flag |
| D | Direction |
| I | Interrupts Enabled |
| T | Trap Flag |
| S | Sign of operation |
| Z | Zero Flag |
| A | Auxiliary Carry |
| P | Parity Flag |
| C | Carry Flag |

As an example, when one compares two numbers and wishes to Jump to the location called Label in the program the operation which is in fact performed is the subtraction of the two numbers and a Jump is executed if the Zero Flag is set (i.e. Z=1). You may compare this to a GO TO statement in FORTRAN or Basic.

---

**12.6** List two advantages of digital signal processing over analog signal processing.

## Solution:

1. Digital signals are less subject to noise, since one only needs to discriminate between two voltages

2. Digital signals are directly compatible with digital computers, and can therefore be easily stored on a disk, or exchanged between computers. Thus, digital signals are intrinsically more portable than analog signals.

---

**12.7** Discuss the role of a multiplexer in a data acquisition system.

**Solution:**

It sequentially switches a set of analog inputs to the system input.

---

**12.8** Discuss the purpose of using sample-and-hold circuits in data acquisition systems.

**Solution:**

A sample-and-hold circuit "freezes" (holds) the value of a signal at the input to an ADC to allow the ADC to convert the analog signal without its changing. The sample-and-hold circuit is necessary because of the finite conversion time of the ADC

---

**12.9** The unsigned decimal number $158_{10}$ is input to an 8-bit DAC. Given that $R_f = \dfrac{R_0}{255}$, logic zero corresponds to 0 volts and logic one corresponds to 10 volts,

a) What is the output of the DAC?

b) What is the maximum voltage that can be output from the DAC?

c) What is the resolution over the range 0 to 10 volts?

d) Find the number of bits required in the DAC if an improved resolution of 3 mV required.

**Solution:**

The only change between problems 13.13 and 13.14 is in part a.

a)     $v_a = -10(\frac{1}{255})(158) = -6.196$ V

b)     $(v_a)_{max} = -10(\frac{1}{255})(255) = -10$ V

c)     $\delta v = 10(\frac{1}{255}) = 39.2$ mV

d)   $n \geq \dfrac{\log(\frac{|(v_a)_{max} - (v_a)_{min}|}{\delta v} + 1)}{\log 2}$

or

$$n \geq 11.703$$

Therefore, we choose $n = 12$.

---

**12.10**   The unsigned decimal number $261_{10}$ is input to a 12-bit DAC . Given that

$R_f = \dfrac{R_0}{4095}$ and logic one corresponds to 10 volts, and logic zero to 0 volts,

a) What is the output of the DAC?

b) What is the maximum voltage that can be output from the DAC?

c) What is the resolution over the 0 to 10 volts?

d) Find the number of bits required in the DAC if an improved resolution of 1 mV required.

**Solution:**

a)     $v_a = -10(\frac{1}{4095})(261) = -0.6374$ V

b)     $(v_a)_{max} = -10(\frac{1}{4095})(4095) = -10$ V

c)     $\delta v = 10(\frac{1}{4095}) = 2.44$ mV

d)   $n \geq \dfrac{\log(\frac{|(v_a)_{max} - (v_a)_{min}|}{\delta v} + 1)}{\log 2}$

or          $n \geq 13.288$

Therefore, we choose $n = 14$.

**12.11** The unsigned decimal number $261_{10}$ is input to a 12-bit DAC. Given that $R_f = \dfrac{R_0}{4095}$ and logic one corresponds to 3.6 volts, and logic zero to 0 volts,

 a) What is the output of the DAC?

 b) What is the maximum voltage that can be output from the DAC?

 c) What is the resolution over the 0 to 10 volts?

 d) Find the number of bits required in the DAC if an improved resolution of 1 mV required.

**Solution:**

a) $\quad v_a = -3.6(\dfrac{1}{4095})(261) = -0.2295$ V

b) $\quad (v_a)_{max} = -3.6(\dfrac{1}{4095})(4095) = -3.6$ V

c) $\quad \delta v = 3.6(\dfrac{1}{4095}) = 0.88$ mV

d) $\quad n \geq \dfrac{\log(\dfrac{|\,(v_a)_{max} - (v_a)_{min}\,|}{\delta v} + 1)}{\log 2}$

or $\qquad\qquad\qquad n \geq 11.814$

Therefore, we choose a 12-bit ADC.

---

**12.12** The unsigned decimal number $345_{10}$ is input to a 12-bit DAC. Given that $R_f = \dfrac{R_0}{4095}$, logic one corresponds to 4.5 volts, and logic zero to 0 volts,

 a) What is the output of the DAC?

 b) What is the maximum voltage that can be output from the DAC?

 c) What is the resolution over the range 0 to 10 volts?

 d) Find the number of bits required in the DAC if an improved resolution of .5 mV required.

## Solution:

a) $v_a = -4.5(\frac{1}{4095})(345) = -0.379$ V

b) $(v_a)_{max} = -4.5(\frac{1}{4095})(4095) = -4.5$ V

c) $\delta v = 4.5(\frac{1}{4095}) = 0.001$mV

d) $n \geq \dfrac{\log(\dfrac{|(v_a)_{max} - (v_a)_{min}|}{\delta v} + 1)}{\log 2}$

or $\qquad\qquad n \geq 13.136$

Therefore, we choose $n = 14$.

---

**12.13** The voltage range of a signal is ±5 volts, and a resolution of .005% of the voltage range is required. How many bits must be in the DAC in order to sense the signal to its full resolution?

## Solution:

The range is 5 - (-5) = 10 V

Thus,

$$\delta v = 10(.005\%) = 0.5 \text{ mV}.$$

$$n \geq \frac{\log(\dfrac{10}{0.5 \text{ mv}} + 1)}{\log 2} = 14.29$$

Choose n = 15

---

**12.14**  What is the maximum sampling time interval to ensure that the Nyquist principle is satisfied when a 75-Hz signal is being sampled?

**Solution:**

$$T_s \leq \frac{1}{150} = 6.67 \text{ ms}$$

---

**12.15**  Discuss the advantages of digital data transmission over analog data transmission.

**Solution:**

The advantage of digital data transmission over analog data transmission is discussed in Section 13.5, p. 716, first paragraph.

---

**12.16**  Explain briefly the concept of interfacing as it is related to instrumentation systems.

**Solution:**

In an instrumentation system, one of the most common requirements is to provide an interface between the analog signals measured by the sensors and a digital computer. This interface usually consists of an ADC, which converts the analog voltages into an equivalent digital representation. The digital signals corresponding to each sampled analog value must then be transmitted to the digital computer. This is often accomplished by means of a bus using a communications protocol; the most common are the IEEE-488 protocol and the RS-232 protocol. These permit transmission of the digital data to the computer in an orderly fashion. The details of such interface protocols are discussed in Section 13.5.

---

**12.17**  Define the following terms.

    a) Handshaking

    b) Baud rate

**Solution:**

a)  Handshaking is defined on p. 716, last paragraph.

b)  Baud is defined in Section 13.5, p. 722, first paragraph.

---

**12.18**  Explain the method of handshaking used in serial transmission.

**Solution:**

Handshaking for serial transmission (RS232) is discussed in section 13.5, on pp. 721-725.

**12.19**  Explain why synchronous data transmission can be faster than asynchronous data transmission.

**Solution:**

Please refer to Section 13.5, p. 716, last paragraph.

---

**12.20**  Discuss the role of DTE and DCE in the RS 232 standard.  Give examples of common DTE and DCE.

**Solution:**

DTE stands for data terminal equipment, and consists of all the devices that can transmit or receive data.  Examples of DTE are computers and digital instruments.  DCE stands for data communication equipment, and consists of those devices used to actually transmit the digital data.  The most common example is the modem (see p. 723).

---

# Section 13:   Principles of Electromechanics

---

**13.1**   An iron core inductor has the following characteristic:

$$\lambda = i - \frac{i^3}{3} + \frac{i^5}{5}; \qquad -1\,A < i < 1\,A$$

a) Determine the energy and the incremental inductance for $i = 0.75$ A.

b) Given that the coil resistance is 1 $\Omega$ and that $i(t) = \sin t$, determine the voltage across the terminals of the inductor.

**Solution:**

The co-energy is computed as follows:

$$W_m' = \int_0^{0.75} (i - \frac{i^3}{3} + \frac{i^5}{5})\,di =$$

$$= 0.2608 \text{ J}$$

(a)  The energy is then computed from

$$W_m = i\lambda - W_m' = 0.2318 \text{ J}$$

and the incremental inductance is:

$$L_\Delta = \frac{d\lambda}{di}\Big|_{i=0.75} = 0.754 \text{ H}$$

(b)  $v = iR + \dfrac{d\lambda}{dt}$   with $R = 1\ \Omega$ and

$i = \sin(t)$

$$\frac{d\lambda}{dt} = \frac{d}{dt}(\sin(t) - \frac{\sin^3 t}{3} + \frac{\sin^5 t}{5})$$

$$= \cos(t) - \sin^2 t\,\cos(t) + \sin^4 t\,\cos(t)$$

$$v = \sin(t) + \cos(t)(1 - \sin^2 t + \sin^4 t)$$

---

**13.2**  An iron core inductor has the following characteristic:

$$\lambda = i - \frac{i^2}{6} + \frac{i^3}{120} \qquad -2\,A < i < 2\,amps$$

a) Determine the energy, coenergy, and the incremental inductance for $i = 1$ amp.

b) Given that the coil resistance is $1\ \Omega$ and that $i(t) = \sin t$, determine the voltage across the terminals on the inductor.

**Solution:**

(a) Given

$$\lambda = i - \frac{i^2}{6} + \frac{i^3}{120} \qquad for -2\,A < i < 2\,A$$

we compute the co-energy:

$$W_m' = \int_0^1 (i - \frac{i^2}{6} + \frac{i^3}{120})\ di$$

$$= 0.4465\ J$$

The energy can be computed from the co-energy using the relation

$$W_m = i\lambda - W_m'$$

$$= 0.8417 - 0.4465 = 0.3952\ J$$

$$L_\Delta = \frac{d\lambda}{di}\Big|_{i=1} =$$

$$= 1 - \frac{i}{3} + \frac{i^2}{40}\Big|_{i=1}$$

$$= 0.6917\ H$$

(b) $\lambda = \sin(t) - \dfrac{\sin^2 t}{6} + \dfrac{\sin^3 t}{120}$

$$\frac{d\lambda}{dt} = \cos(t) - \cos(t)\frac{\sin(t)}{3} + \cos(t)\frac{\sin^2(t)}{40}$$

Therefore,

$$v_L(t) = \sin(t) + \cos(t)\{1 - \frac{\sin(t)}{3} + \frac{\sin^2(t)}{40}\}$$

**13.3** With reference to Figure 15.17, suppose that $A = A_g$ 0.01 m², $\delta$ = 0.005 m, N = 1000 turns. Find the current if a flux of 0.01 Wb in the air gap is required, neglecting iron reluctance.

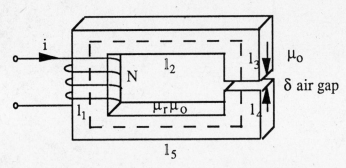

**Figure 15.17 Magnetic circuit with air gap**

**Solution:**

For a gap area equal to

$$A = A_g = 0.01 \text{ m}^2$$

$$\mathbb{R}_g = \frac{0.005}{(4\pi \times 10^{-7})(0.01)} =$$

$$= 398 \times 10^3 \text{ H}^{-1}$$

From $\phi_g = \dfrac{Ni}{\mathbb{R}_g}$, we can compute the current

$$i = \frac{\phi_g \mathbb{R}_g}{N} =$$

$$= \frac{0.01(398 \times 10^3)}{1000} = 3.98 \text{ A}$$

---

**13.4** With reference to Figure 15.17, suppose that $A = A_g$ 0.01 m², $\delta$ = 0.005 m, N = 1000 turns. Find the current that will produce a flux density in the air gap of 1.5 T.

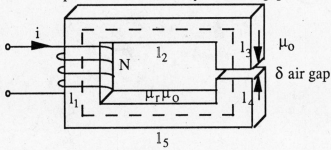

**Figure 15.17 Magnetic circuit with air gap**

**Solution:**

For a flux density of $B_g = 1.5$ T, we can compute the flux, knowing the cross-sectional area:

$$\phi_g = 1.5 A_g = 0.015 \text{ Wb}$$

Therefore,

$$i = \frac{0.015(398 \times 10^3)}{1000} = 5.97 \text{ A}$$

---

**13.5**  The cylindrical solenoid shown in Figure P14.4 carries a constant current of 5 A.  Its armature is allowed to more slowly from x = 5 mm to 2 mm.

Find (a)  the change in energy stored in the magnetic field

(b)  the change in the magnetic coenergy

(c)  the change in the energy supplied by the source

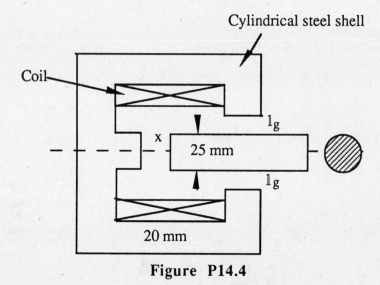

**Figure  P14.4**

**Solution:**

a) The stored energy in a magnetic field is

$$W_m = \frac{N^2 i^2}{2 \mathbb{R}}$$

while the change in reluctance caused by a movement from 2 mm to 5 mm is:

$$\Delta \mathbb{R} = 1621 \times 10^6 (3 \times 10^{-3})$$

Therefore, the change in stored energy is:

$$\Delta W_m = \frac{N^2 i^2}{2 \Delta R} = \frac{100^2 \times 7.5^2}{2 \times 1621 \times 10^6 \times 3 \times 10^{-3}}$$

$$= 0.0578 \ J$$

b) $\Delta W'_m = \Delta W_m = 0.0578 \ J$

c) The change in the energy supplied by the source will be equal to the change in energy stored in a conservative system.

---

**13.6**  For the circuit shown in Figure P14.7:

    a) Determine the reluctance values and show the magnetic circuit assuming that $\mu$ = $\mu_r \mu_o$ and that $\mu_o = 4\pi \times 10^{-7}$ H/m.

    b) Determine the inductance of the device shown above.

    c) If an air gap of length 0.1 mm is cut in the magnetic structure directly across from the coil, what is the new value of inductance?

    <u>Neglect leakage flux and fringing effects.</u>

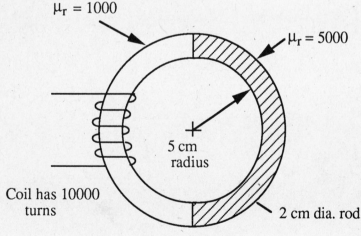

**Figure  P14.7**

**Solution:**

The length of the two semi-circular legs is:

$$l_1 = l_2 = \frac{\pi d}{2} = \pi 6 \times 10^{-2} = 0.188 \text{ m}$$

The cross-sectional areas are:

$$A_1 = A_2 = \pi r^2 = \pi(1 \times 10^{-2})^2 = 3.14 \times 10^{-4} \text{ m}^2$$

(a) The circuit is shown below

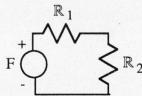

$$\mathbb{R}_1 = \frac{0.188}{1000(4\pi \times 10\text{-}7)3.14 \times 10^{-4}} = 4.76 \times 10^5 \text{ H}^{-1}$$

$$\mathbb{R}_2 = 9.52 \times 10^4 \text{ H}^{-1}$$

$$\mathbb{R}_T = \mathbb{R}_1 + \mathbb{R}_2 = 5.71 \times 10^5 \text{ H}^{-1}$$

(b) $L = \dfrac{N^2}{\mathbb{R}_T} = \dfrac{(10^4)^2}{5.71 \times 10^5} = 175.1 \text{ H}$

(c) The reluctance of the air gap is

$$\mathbb{R}_g = \frac{0.0001}{4\pi \times 10^{-7} \times 3.14 \times 10^{-4}} = 2.53 \times 10^5$$

$$\mathbb{R}_T = \mathbb{R}_1 + \mathbb{R}_2 + \mathbb{R}_g = 8.24 \times 10^5 \text{ H}^{-1}$$

$$L = \frac{N^2}{\mathbb{R}_T} = 121 \text{ H}$$

**13.7**

a) For the electromagnet shown in Figure P14.8, assume the permeability of the magnetic material is infinite so that the reluctance is all due to the air-gap. For a force f = 2000 N and a DC excitation current I = 25 A ($A_g = 20$ cm$^2$) determine the number of turns N required to hold the air-gap (i.e. $x$) at 0.1 cm.

b) What peak value of AC current would be required to generate the same average force ?

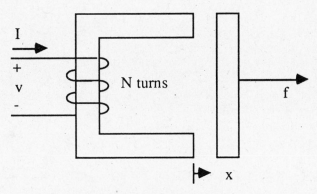

**Figure P14.8**

**Solution:**

(a) From $\mathbb{R}_g(x) = \dfrac{x}{\mu_0 A_g}$, we have

$$\mathbb{R}_T(x) = 2\mathbb{R}_g(x) = \frac{2x}{\mu_0 A_g} = \frac{2x}{4\pi \times 10^{-7}(20 \times 10^{-4})} = 7.96 \times 10^8 x$$

The force in the air gap is:

$$f = \frac{i^2}{2} \frac{N^2}{\mathbb{R}_T^2} \frac{d\mathbb{R}_T}{dx}$$

Therefore,

$$N^2 = \frac{2\mathbb{R}_T^2 f}{i^2 \dfrac{d\mathbb{R}_T}{dx}}\Big|_{x = 0.1 \text{ cm}} = 5094.4$$

and

$$N = 71 \text{ turns}$$

(b) The rms value of the current should be 25 A and the peak value is

$$i_{peak} = 35.36 \text{ A}$$

**13.8**   For the DC relay shown in Figure P14.9, the resistance of the winding is R = 25 Ω, N = 100 turns, $x$ = 0.1 cm., and the cross section area of the gap is 0.01 m². At this position the spring force is 2 N. Assume the permeability of the magnetic material is infinite and that $\mu = \mu_0 = 4\pi \times 10^{-7}$ H/m. for the air gap.

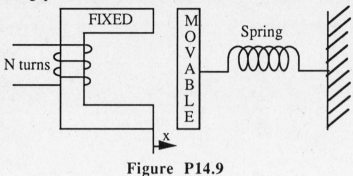

**Figure P14.9**

a) What current is required to hold the movable piece in this position and what is the input voltage to get this current?

b) What is the inductance, L(x), and the energy stored in the magnetic field $W_m$ at this position?

**Solution:**

The variable gap reluctance is:

$$\mathbb{R}_T(x) = \frac{2x}{\mu_0 A g} = 1.59 \times 10^5 \, x$$

(a)  Thus, the force in the air gap is:

$$- f = - \frac{i^2}{2} \frac{N^2}{(\mathbb{R}(x))^2} \frac{d\mathbb{R}(x)}{dx} = - i^2 \frac{100^2 \times 1.59 \times 10^5}{2 \times (1.59 \times 10^5)^2 x^2} \, N$$

This force will oppose the 2-N spring force, as illustrated in the equation below:

$$- 2 = - i^2(3.14 \times 10^{-2}) \frac{1}{x^2} = - 3.14 \times 10^4 \, i^2$$

The current can now be computed to be:

$$i = \sqrt{\frac{2}{3.14 \times 10^4}} = 7.98 \text{ mA}$$

$$V = R \, i = 0.2 \text{ V}$$

(b)  The inductance is :

$$L(x) = \frac{N^2}{\mathbb{R}_T(x)} =$$

$$= \frac{6.28 \times 10^{-2}}{x} \Big|_{x=0.001} = 62.83 \text{ H}$$

and the stored energy is equal to:

$$W_m = L(x) \frac{i^2}{2} = 2.002 \times 10^{-3} \text{ J}$$

---

**13.9**  A core is shown in Figure P14.10, with $\mu_r = 1200$, $N = 100$.  Find

    (a)  the flux in each of the legs

    (b)  the flux density in each of the legs

        Neglect fringing effects at the air gap.

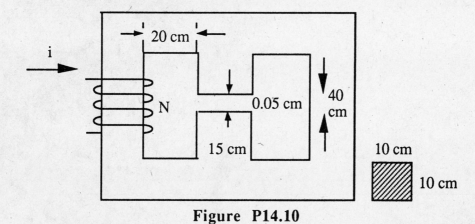

**Figure  P14.10**

## Solution:

The analogous circuit is shown below:

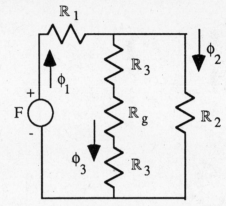

We have

$$\mathbb{R}_1 = \frac{0.3 + 0.3 + 0.5}{\mu_0 \mu_r A} = 7.29 \times 10^4 = \mathbb{R}_2$$

$$\mathbb{R}_3 \approx \frac{0.25}{\mu_0 \mu_r (1.5 \times 10^{-2})} = 1.105 \times 10^4$$

$$\mathbb{R}_g = \frac{0.05 \times 10^{-2}}{\mu_0 (1.5 \times 10^{-2})} = 2.65 \times 10^4$$

b) $\phi_1 = \dfrac{F}{\mathbb{R}_T}$,  $F = Ni = 10$ A-t

$$\phi_1 = \frac{F}{(2\mathbb{R}_3 + \mathbb{R}_g) \| \mathbb{R}_2 + \mathbb{R}_1}$$
$$= 9.8 \times 10^{-5} \text{ Wb}$$

$$\phi_2 = \frac{\phi_1 (2\mathbb{R}_3 + \mathbb{R}_g)}{2\mathbb{R}_3 + \mathbb{R}_g + \mathbb{R}_2} = 3.92 \times 10^{-5} \text{ Wb}$$

$$\phi_3 = \phi_1 - \phi_2 = 5.88 \times 10^{-5} \text{ Wb}$$

b) $B_1 = \dfrac{\phi_1}{A} = 9.8 \times 10^{-3}$ T

$$B_2 = \frac{\phi_2}{A} = 3.92 \times 10^{-3} \text{ T}$$

$$B_3 = \frac{\phi_3}{1.5 \times 10^{-2}} = 3.92 \times 10^{-3} \text{ T}$$

**13.10**   For the magnetic circuit shown in Figure P14.11, the self inductance for the two coils is: $L_{11} = L_{22} = 5 + (2x)^{-1}$ mH , and the mutual inductance between the two coils is: $M_{12} = M_{21} = (4x)^{-1}$ mH, where x is in meters.

  a)  For $i_1 = 7.5$ A and $i_2 = 0$ what is the force between the coils for $x = 0.01$ meters?

  b)  For the same currents as in a) determine the voltage across the movable coil at x $= 0.01$ meters when the movable coil is moving in the $x$ direction at a constant speed of 10 meters/sec.

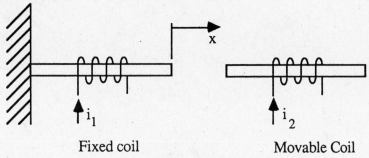

Fixed coil                    Movable Coil

**Figure  P14.11**

**Solution:**

(a) $W_m' = \frac{1}{2}L_{11}i_1^2 + \frac{1}{2}L_{22}i_2^2 + M_{12}i_1i_2$

Since $i_2 = 0$, this reduces to

$W_m' = \frac{1}{2}L_{11}i_1^2$

$= 140.63 + \dfrac{14.06 \times 10^{-3}}{x}$

$$f = -\frac{\partial W_m'}{\partial x} = \frac{14.06 \times 10^{-3}}{x^2}$$

For $x = 0.01$, the force is

$$f = \frac{14.06 \times 10^{-3}}{(0.01)^2} = 140.6 \text{ N}$$

(b) $\qquad v_2 = \dfrac{d\lambda_2}{dt} = \dfrac{\partial \lambda_2}{\partial i}\dfrac{di}{dt} + \dfrac{\partial \lambda_2}{\partial x}\dfrac{dx}{dt}$

$$\lambda_2 = M_{21}i_1 + L_{22}i_2 = \frac{10^{-3}}{4x} \times 7.5$$

Since $i_1 = 7.5$ A and $i_2 = 0$, $\dfrac{di}{dt} = 0$

$\dfrac{dx}{dt} = 10$, therefore,

$$v_2 = -\frac{10^{-3} \times 7.5}{4 \times (0.01)^2} \times 10 = -187.5 \text{ V}$$

---

**13.11**   An electromechanical system is shown in Figure P14.13.  Find the differential equations describing the system after the switch is closed.  Neglect fringing and leakage and the friction.

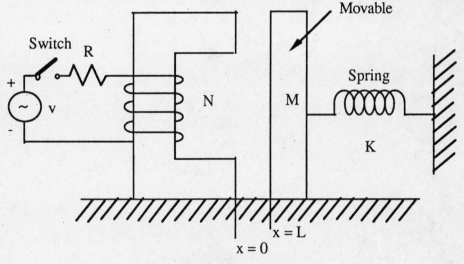

**Figure  P14.13**

**Solution:**

The equation for the electrical system is:

$v = i\,R + L(x)\,\dfrac{di}{dt}$ , where

$$L(x) = \frac{N^2}{\mathbb{R}_T(x)} = \frac{N^2 \mu_0 A}{2x}$$

The equation for the mechanical system is:

$F_m = m\,\dfrac{d^2x}{dt^2} + k\,x$, where $F_m$ is the magnetic pull force. To calculate this force we use equation 14.41,

$F_m = -\dfrac{dW_m}{dx}$ , where $W_m$ is the energy stored in the magnetic field. Let $\mathbb{F}$ and $\mathbb{R}$ be the magnetomotive force acting on the structure and its reluctance, respectively; then

$$W_m = \frac{\phi^2 \mathbb{R}(x)}{2} = \frac{\mathbb{F}^2}{2\,\mathbb{R}(x)} = \frac{N^2\,i^2\,\mu\,A}{4x}$$

and $\quad F_m = -\dfrac{dW_m}{dx} = \dfrac{N^2\,i^2\,\mu\,A}{4x^2}$ .

Finally, the differential equations governing the system are:

$$v = i\,R + \frac{N^2 \mu_0 A}{2x}\,\frac{di}{dt}$$

$$m\,\frac{d^2x}{dt^2} + k\,x = \frac{N^2\,i^2\,\mu\,A}{4x^2}\ .$$

This system of equations could be solved using a numerical simulation, since it is nonlinear.

**13.12**  A relay is shown in Figure P14.14.  Find the differential equations describing the system.

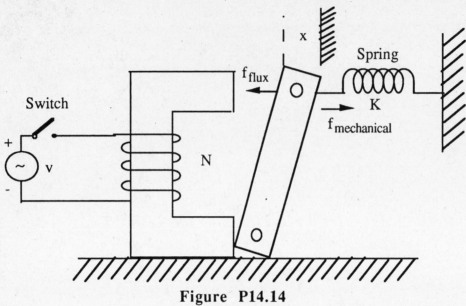

**Figure  P14.14**

**Solution:**

The equation for the electrical system is:

$v = i R + L(x) \dfrac{di}{dt}$ , where

$$L(x) = \frac{N^2}{\mathbb{R}_T(x)} = \frac{N^2 \mu_0 A}{2x}$$

The equation for the mechanical system is:

$F_m = m \dfrac{d^2x}{dt^2} + k\,x$, where $F_m$ is the magnetic pull force. To calculate this force we use equation 14.41,

$F_m = - \dfrac{dW_m}{dx}$ , where $W_m$ is the energy stored in the magnetic field. Let $\mathbb{F}$ and $\mathbb{R}$ be the magnetomotive force acting on the structure and its reluctance, respectively; then

$$W_m = \frac{\phi^2 \mathbb{R}(x)}{2} = \frac{\mathbb{F}^2}{2\,\mathbb{R}(x)} = \frac{N^2 i^2 \mu A}{4x}$$

and $\qquad F_m = - \dfrac{dW_m}{dx} = \dfrac{N^2 i^2 \mu A}{4x^2}$ .

Finally, the differential equations governing the system are:

$$v = i R + \frac{N^2 \mu_0 A}{2x} \frac{di}{dt}$$

$$m \frac{d^2x}{dt^2} + k\,x = \frac{N^2 i^2 \mu A}{4x^2} .$$

This system of equations could be solved using a numerical simulation, since it is nonlinear.

---

**13.13** The model of on electromechanical conversion system is shown in Figure P14.15. Find the differential equations describing the system.

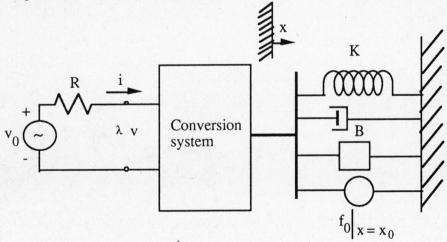

**Figure P14.15**

**Solution:**

Figure P14.15 illustrates the interface between the electrical and mechanical parts of a system. The coupling between the two is assumed to be of magnetic origin. The force balance equation for the mechanical side of the system is:

$$f_0 = m \frac{d^2x}{dt^2} + d \frac{dx}{dt} + k + K_m i$$

where $K_m$ is the conversion constant from the *Bli* law. If we assume the system to be lossless, and to have inductance L, we can also obtain the equations of the electrical side:

$$v_0 = iR + L \frac{di}{dt} + K_m \frac{dx}{dt}$$

These differential equations are coupled in that the motion of the mechanical system gives rise to an electromotive force (Blu law), and the current in the electrical system gives rise to a mechanical force (Bli law). In this case we have simply modeled the magnetic coupling between systems as being linear and conservative (thus the identical coefficients $K_m$). Could you sketch an example of a physical device that would behave according to the equations just derived?

---

**13.14**   An air gap inductor shown in Figure P14.16 has a height, *h*, of 2.5 cm, width, *w*, of 2.5 cm, air gap dimension lg = 0.1 cm, N = 400 turns, and is excited by a DC current I = 5 A. Determine the force of attraction between sides of the gap, neglecting the reluctance of the core, leakage and fringing.

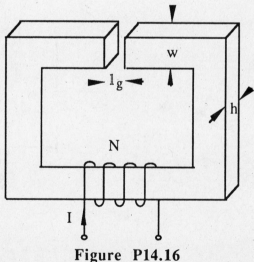

**Figure  P14.16**

**Solution:**

$$\mathbb{R}_T = \frac{x}{4\pi \times 10^{-7}(0.025)^2} = 1.27 \times 10^9 x$$

$$f = -\frac{N^2}{2} \frac{i^2}{\mathbb{R}_T^2} \frac{d\mathbb{R}_T}{dx} = -1.57 \times 10^{-3} \frac{1}{x^2}$$

For $x = 0.001$ m, $f = -1574.8$ N

The minus sign indicates that the force is in a direction that would decrease the gap.

---

**13.15** A wire of length 20 cm vibrates in one direction in a constant magnetic field with a flux density of 0.1 T shown in Figure P14.17. The position of the wire as a function of time is given by $x(t) = 0.1 \sin 10t$ m. Find the induced emf across the length of wire as a function of time.

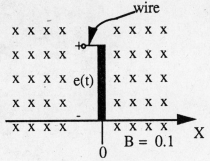

**Figure P14.17**

**Solution:**

From $e = Blu$, we have

$$e(t) = Bl\frac{dx}{dt} = 0.02\cos(10t) \text{ V}$$

---

**13.16**  A wire vibrating in the magnetic field induced a time varying emf of
$$e_1(t) = 0.02 \cos 10t.$$
A second wire is placed in the same magnetic field but has a length of 0.1 meters as shown in Figure P14.18.  The position of this wire is given by $x(t) = 1 - 0.1 \sin 10t$.  Find the induced emf $e(t)$ defined by the difference in emfs $e_1(t)$ and $e_2(t)$.

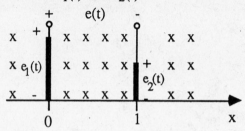

**Figure  P14.18**

**Solution:**

We have

$e(t) = e_1(t) - e_2(t)$

$e_2(t) = (0.1)(0.1)(-1 \cos 10t)$ V

$e(t) = 0.02\cos(10t) + 0.01\cos(10t)$

$\quad = 0.03\cos(10t)$ V

**13.17** A conducting bar shown in Figure 15.47 is carrying 4 A current in the presence of a magnetic field. B = 0.3 Wb/m². Find the magnitude and direction of the force induced on the conducting bar.

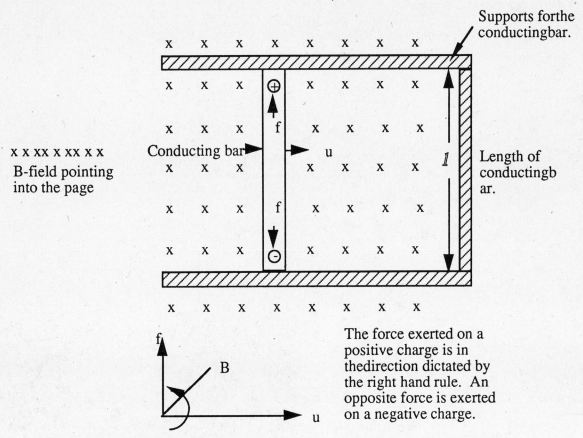

**Figure 15.47 A simple electromechanical motion transducer**

**Solution:**

$$f = Bli = 0.3 \times l \times 4 = 1.2l \text{ N}$$

Force will be to the left if current flows upward.

**13.18** A wire show in Figure P14.20 is moving in the presence of a magnetic field B = 0.4 Wb/m². Find the magnitude and direction of the induced voltage in the wire.

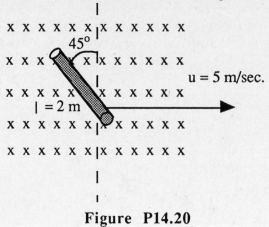

**Figure P14.20**

**Solution:**

$$e = Blu \cos 45° = 2.83 \ V$$

# Section 14:   Introduction to Electric Machines

**14.1**   A DC shunt motor has a no load saturation characteristic given by

$$E_b = \frac{300\, i_f}{1 + i_f} \quad \text{(at 1000 rev/min)}$$

where E is the open circuit voltage and $i_f$ the shunt field current.  The armature resistance is 0.12 ohms and the shunt field resistance is 247 ohms.  The motor is operating from a 230 volt DC source.

a) If the load is adjusted to make the armature current 80 amps, find the speed in revolutions per minute.

b)  If the no load line current is 3 amps, determine the value of $T_{shaft}$ necessary to make the armature current 80 amps, (assume $T_{start} = 0$).

## Solution:

$$I_f = \frac{230}{247} = 0.931 \text{ A}$$

(a) For $n_m = 1000$ rev/min,

$$\omega_r = \frac{2\pi}{60} \times 1000 = 104.7 \text{ rad/sec}$$

Therefore, at $\omega_m = 104.7$ rad/sec,

$$E_b = \frac{300 \times 0.931}{1 + 0.931} = 144.64 \text{ V}$$

From $E_b = k_a\phi\omega_m = 144.64$ V, we have

$$k_a\phi = \frac{144.64}{104.7} = 1.38 \text{ VS/rad}$$

For $i_a = 80$ A, we have

$$E_b = 230 - 80 \times 0.12 = 220.4 \text{ V}$$

and

$$\omega_m = \frac{220.4}{1.38} = 159.7 \text{ rad/sec}$$

$$= 1525 \text{ rev/min}$$

(b) The shaft torque is given by:

$$T_{shaft} = T - T_{SL} - T_r$$

where

$T_{SL}$ is the stray load loss

$T_r$ is the rotational loss = no-load torque

We have

$$i_a = 3 - 0.931 = 2.069 \text{ A}$$
$$E_b = 230 - 2.069 \times 0.12 = 229.8 \text{ V}$$
$$\omega_m = \frac{229.8}{1.38} = 166.5 \text{ rad/sec}$$

From $P_r = E_b i_a = 475.5$ W, we have

$$T_r = \frac{P_r}{\omega_m} = 2.86 \text{ N-m}$$

$T_{shaft}$ at full load is found to be:

$$T_{shaft} = k_a\phi i_a - T_r = 107.5 \text{ N-m}$$

**14.2**  A separately excited generator produces 400 V at 1500 rev/min.  Find the induced emf if the speed changes to 2000 rev/min.  Assume flux is constant.

**Solution:**

If $\phi$ is kept constant, $E_b = k_a \phi \omega_m$ will be proportional to $\omega_m$.

For $E_b = 400$ V at 1500 rev/min (157.1 rad/sec), we have:

$E_b = 533.3$ V at 2000 rev/min (209.44 rad/sec).

---

**14.3**  The voltage regulation of a 220 V shunt generator is 10%.  Find the terminal voltage when the load is reduced to zero.

**Solution:**

From the definition of voltage regulation,

$$\text{Voltage reg.} = \frac{v_L - 220}{220} = 0.1$$

$v_L$ at no load is 242 V

---

**14.4**  A  400 V, 100 kW, long shunt compound generator has a series field resistance of 0.04 $\Omega$, a shunt field resistance of 100 $\Omega$ and an armature resistance of 0.02 $\Omega$.  The rated current is delivered at rated speed of 100 rev/min.

Find  (a)  armature current

      (b)  armature voltage

**Solution:**

The circuit is shown below:

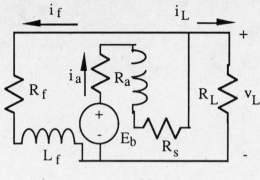

$$v_L = 400 \text{ V}$$

$$i_L = \frac{100 \times 10^3}{400} = 250 \text{ A}$$

$$n_m = 100 \text{ rev/min}; \quad \omega_m = 10.47 \text{ rad/sec}$$

(a) $i_f = \frac{400}{100} = 4 \text{ A}$

$i_a = i_L + i_f = 254 \text{ A}$

(b) $E_b = v_L + i_a(R_a + R_S) =$

$\qquad = 400 + 254 \times 0.06 = 415.24 \text{ V}$

---

**14.5**   The open-circuit characteristic for a DC shunt generator operating at 1500 rev/min is given by

$$E = 367 I_f / (0.233 + I_f)$$

where E is the open-circuit voltage and $I_f$ is the shunt field current.  The other parameters are

$$R_a = 0.14 \ \Omega \quad R_f = 459 \ \Omega.$$

   a)  Determine the no-load terminal voltage when this machine is operating as a self-    excited shunt generator.

   b)  What is the terminal voltage for an armature current of 50 A?

   c)  What is the terminal voltage for the compound generator for an armature current of   50 amps?

**Solution:**

Let $E_b = \dfrac{367 I_f}{0.233 + I_f}$ with $R_a = 0.14 \ \Omega$ and

$R_f = 459 \ \Omega$.

(a) Under no load conditions,

$$I_f = I_a, \quad E_b = I_f(R_a + R_f)$$

Therefore,

$$\frac{367 I_f}{0.233 + I_f} = 459.14 I_f$$

$$367 I_f = 106.98 I_f + 459.14 \ I_f^2$$

Thus,

$$I_f = 0.566 \ A$$

and $\quad v_L = E_b - i_f(0.14) = 259.8 \ V$

(b) $i_a = 50$ A when the machine is loaded

From $E_b - i_a R_a = v_L$, we have

$$v_L = 259.9 - 50(0.14) = 252.9 \ V$$

---

**14.6** A 230 volt DC shunt motor has $R_a = 0.05$ ohms and $R_f = 75$ ohms.

a) At no load the motor draws 7 amps from the line and runs at a speed of 1120 rev/min. Determine the copper losses and the rotational loss (assuming the stray losses to be negligible) for this machine operating at no load.

b) The machine is loaded such that the line current becomes 46 amps. What is the speed (in rev/min) at this load, and what is the developed torque and the shaft torque?

**Solution:**

(a) $i_f = \dfrac{230}{75} = 3.07$ A, $i_a = i_S - i_f = 3.93$ A

$E_b = \omega_m G_m i_f = v_S - i_a R_a$

$\quad = 230 - 3.93 \times 0.05 = 229.8$ V

$$\omega_m = 117.3 \text{ rad/sec}$$

$G_m = \dfrac{229.8}{117.3 \times 3.07} = 0.638$ VS/A rad

The copper loss is

$$P_{copper} = i_a^2 R_a + i_f^2 R_f = 707.6 \text{ W}$$
$$P_{in} = 7 \times 230 = 1610 \text{ W}; \quad P_{out} = 0$$

Therefore, the rotation loss is

$$P_{rot} = 1610 - 707.6 = 902.36 \text{ W}$$
$$T_{rot} = P_{rot}/\omega_m = 7.7 \text{ N-m}$$

(b) $i_S = 46$ A $\quad i_a = 42.93$ A $\quad E_b = 227.9$ V

Assuming $G_m$ is constant, we have

$$\omega_m = \dfrac{227.9}{0.638 \times 3.07} = 116.4 \text{ rad/sec}$$

The developed power is

$$P = E_b i_a = 9.784 \text{ kW}$$
$$T = P/\omega_m = 84.1 \text{ N-m}$$

Assuming $T_{rot}$ is constant, we have

$$T_{SH} = 84.1 - 7.7 = 76.4 \text{ N-m}$$

---

**14.7** A DC shunt motor has its no load saturation characteristic given by

$$E = \dfrac{1020 I_f}{0.9 + I_f} \text{ at 1200 rev/min}$$

where $E_b$ is the generated voltage and $I_f$ is the shunt field current. The armature resistance is 0.48 ohms, the field resistance is 368 ohms, and the supply line current 50A. at a terminal voltage of 460 volts.

a) What is the speed (in rev/min)?

b) What is the developed torque?

**Solution:**

$$I_f = \frac{460}{368} = 1.25 \text{ A}$$

At $n_m$ = 1200 rev/min, we have

$$E_b = \frac{1020 \times 1.25}{0.9 + 1.25} = 593 \text{ V}$$

$$i_a = i_S - i_f = 50 - 1.25 = 48.75 \text{ A}$$

and

$$E_b' = 460 - 48.75 \times 0.48 = 436.6 \text{ V}$$

From $\dfrac{E_b'}{E_b} = \dfrac{n_m'}{n_m}$, the speed is:

$$n_m' = \frac{436.6}{593} \times 1200 = 883.5 \text{ rev/min}$$

or $\omega_m = 92.5$ rad/sec

The power developed is

$$P_{dev} = E_b' i_a = 21.28 \text{ kW}$$

The torque is

$$T_{dev} = \frac{P_{dev}}{\omega_m} = 230.1 \text{ N-m}$$

---

**14.8**  A non-salient pole, Y connected, 3-phase, 2-pole synchronous machine has $X_S$ = 1.9 ohms and negligible phase resistance and rotational losses. The field current is adjusted to give an excitation voltage of 380 volts per phase (open circuit voltage when operated as a generator). It is to be connected as a motor to a 254 volt (per phase) line and will drive a load such that $|\delta| = 30°$. Find the armature current and power factor. Is the current capacitive or inductive?

**Solution:**

From the relation $E_b = 380\angle\delta$, $V_S = I_S j1.9 + E_b$, we have

$$I_S = \frac{204\angle 111.57°}{1.9\angle 90°} = 107.5\angle 21.57° \text{ A}$$

$$pf = \cos 21.57° = 0.93 \text{ leading}$$

The current is capacitive.

---

**14.9** A non-salient pole, Y connected, 3-phase, 2-pole synchronous motor is connected to a 220 volt (line-to-line) 60 Hz source. The motor is driving a load such that $T_{load} = 85$ Nt·m. and is known to have total rotational and stray load losses of 942.5 watts. The total stator copper loss at this load is 600 watts. The measured phase parameters are $R_s = 0.02$ ohms and $X_S = 1.269$ ohms.

a) Find $I_s$, $E_b$, $\delta$, power factor and total power into the machine.

b) There are two equally valid solutions for this problem as specified.

Explain why that is possible.

**Solution:**

We have $V_s = 127$ V, $T_{load} = 85$ N-m,

$$\omega_m = 377 \text{ rad/sec}$$

therefore,

$P_{out} = 377 \times 85 = 32045$ W

$P_{in} = P_{out} + P_r + P_{sl} + P_{copper} = 33587.5$ W

On a per-phase basis, $P'_{in} = 11196$ W

From $P_{copper} = \dfrac{600}{3} = I_s^2 R_s$, we calculate

$$I_s = 100A$$

From $P'_{in} = 100 \times 127 \times \cos\theta$, we calculate

$$pf = \cos\theta = 0.8816$$

$\theta = -28.16°$ (assuming an inductive load)

$Z_s = 0.02 + j1.269 = 1.269\angle 89.1° \ \Omega$

$E_b = 128.7\angle -59.5°$ V

$\delta = -59.5°$

**14.10** A synchronous motor with a leading power factor can be used to correct power factor. Assume the power of a load is S' with a lagging power factor angle θ. The apparent power requirement of a synchronous motor is $S'_m$ at a leading power factor angle $θ_m$. The connection is shown in Figure P15.4.

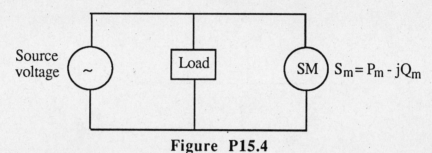

**Figure P15.4**

Find the new power factor and explain how the synchronous motor can be used to improve the power angle.

**Solution:**

For the motor, $S_m = P_m - jQ_m$

For the load, $S_L = P_L + jQ_L$

For the input,

$$S_T = P_m + P_L + j(Q_L - Q_m)$$

The overall phase angle is

$$θ = \tan^{-1}\frac{Q_L - Q_m}{P_L + P_m}$$

By adjusting $i_f$, the magnitude of $Q_m$ can be adjusted to cancel all or part of $Q_L$ in the overall input power to make the pf → 1.

**14.11** Draw the phasor diagram of the synchronous motor of the previous problem as shown in Figure P15.4.

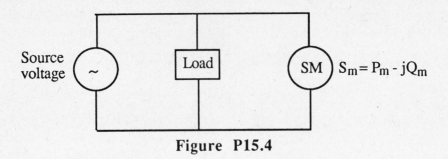

Source voltage ~     Load     SM $S_m = P_m - jQ_m$

**Figure P15.4**

**Solution:**

The diagram is shown below:

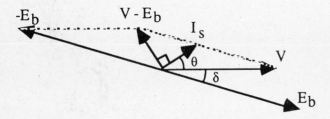

**14.12** A four-pole induction motor operating at 60 Hz has a full-load rotor slip of 5 percent.
Calculate the rotor frequency

    a.     At the instant of starting.

    b.     At full load.

**Solution:**

a) $n_s = \dfrac{120f}{P} = \dfrac{120 \times 60}{4} = 1800$ rev/min

$$f_R = f_s = f = 60 \text{ Hz}$$

b) $f_R$ = slip frequency = $0.05 \times 60 = 3$ Hz

**14.13** Calculate the full-load rotor speed of a four-pole induction motor operating at 60 Hz if the slip is

    a.  5 percent when the line frequency is 60 Hz.

    b.  7 percent when the line frequency is 50 Hz.

**Solution:**

a)  $s = \dfrac{n_s - n}{n_s}$

From  $0.05 = \dfrac{1800 - n}{1800}$, we have

$$n = 1710 \text{ rev/min}$$

b)  $n_s = \dfrac{120 \times 50}{4} = 1500 \text{ rev/min}$

From $0.07 = \dfrac{1500 - n}{1500}$, we have

$$n = 1395 \text{ rev/min}$$

---

**14.14**  A motor will be used to drive an 18 kW load working intermittently. Let the overload capability of a motor, $\lambda$, be the ratio of the maximum intermittent power the motor can produce to the nominal motor power rating. Select one of the following two motors based on overload capability.

Motor 1: $P_e$= 10 kW, $n$ =1460 rpm, $\lambda$=2.5.

Motor 2: $P_e$= 14 kW, $n$ =1460 rpm, $\lambda$=2.8.

**Solution:**

For motor 1, the maximum overload capability is:

$$10 \times 2.5 = 25.0 \text{ kW}$$

The overload capability of motor 2 is:

$$14 \times 2.8 = 39.2 \text{ kW}$$

Assume the starting power is twice rated power. The starting power of motor 1 is:

$$2 \times 10 = 20 \text{ kW}$$

For motor 2:     $2 \times 14 = 28$ kW

Because the motor is working intermittently and both the overload and starting capabilities of motor 1 meet the load requirements, we should choose motor 1 for this application.

---

**14.15**   A 10 hp, 230 V, 3-phase induction motor is marked with code letter G. Its starting kVA per horsepower will be less than 6.3. Find the upper limit of the starting current.

**Solution:**

If the limit is 6.3 kVA/hp and the line voltage is 230 V, then the starting current is limited to

$$i_{max} \leq \frac{6.3 \times 10^4}{\sqrt{3}\,(230)} = 158.1 \text{ A}$$

---

**14.16**   Repeat above problem for the following induction motor.

| MODEL | 18300 J-X | | |
|---|---|---|---|
| **TYPE** | CJ4B | **FRAME** | 324TS |
| **VOLTS** | 230/460 | **ºC AMB.** | 40 |
| | | **INS. CL.** | B |
| **FRT. BRG** | 210SF | **EXT. BRG** | 312SF |
| **SERV FACT** | 1.0 | **OPER INSTR** | C-517 |
| **PHASE     3** | **Hz     60** | **CODE     G** | **WDGS     1** |
| **H.P.** | 40 | | |

| R.P.M. | 3565 | | |
|---|---|---|---|
| AMPS | 106/53 | | |
| NEMA NOM. | EFF | | |
| DUTY | CONT. | NEMA DESIGN | B |

## Solution:

For the machine of Example 15.2, the limit is again 6.3 kVA/hp. The machine is a 40 hp machine. Thus for 230-V line-to-line voltage, we have the maximum current

$$i_{max} \leq \frac{(6.3 \times 10^3)40}{\sqrt{3}\,(230)} = 632.6 \text{ A}$$

---

**14.17** Find the Thèvenin equivalent of the circuit of Figure 16.40.

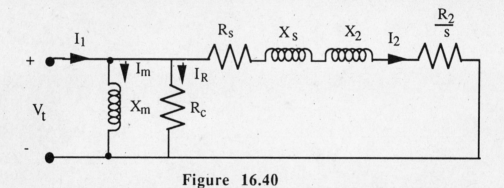

**Figure 16.40**

## Solution:

Assume $R_c$ is an open circuit. Then, the Thèvenin voltage is

$$V_T = V_s \frac{Z_m}{Z_m + Z_s}$$

$$= V_s \frac{jX_m}{R_s + jX_s + jX_m}$$

The magnitude is

$$\text{for } (X_m + X_s) \gg R_s$$

$$V_T = V_s \frac{X_m}{\sqrt{R_s^2 + (X_s + X_m)^2}}$$

$$\approx V_s \frac{X_m}{X_s + X_m}$$

The Thèvenin impedance is

$$Z_T = \frac{Z_s Z_m}{Z_s + Z_m} = R_T + jX_T$$

$$= \frac{jX_m(R_s + jX_s)}{R_s + j(X_s + X_m)}$$

For $X_m \gg X_s$, $(X_m + X_s) \gg R_s$, we have

$$R_T \approx R_s$$

$$X_T \approx X_s$$

---

**14.18** Explain the disadvantages of

    a.    Using a 20 hp motor to drive a 10 hp load continously.

    b.    Using a 5 hp motor to drive a 10 hp load continously.

## Solution:

a) The capacity of the motor is not fully used and the efficiency is low.

b) The motor may be able to supply the required power, but it will overheat and damage will eventually result. It should be remarked that thermal overload is the limiting factor in determining the performance of many electric motors.

---

**14.19** In addition to the factors of rated voltage, rated current, frequency, temperature rise, speed, and duty cycle, give some other considerations affecting the selection of

  a.    Generators.

  b.    Motors.

**Solution:**

Overload capability; starting capability; the altitude at which a machine operates; the load characteristic; cost considerations; working environment.

# Section 15: Special-Purpose Electric Machines

**15.1** From a mechanical standpoint, the brushless motor consists of how many parts?

**Solution:**

The major parts of a DC brushless motor are: (i) the stator with a multiphase winding; (ii) a permanent magnet rotor; and (iii) a rotor position sensor. There will be some other mechanical parts that are required for any application. For example, it must be mounted in some kind of structure (case) and the motor shaft will also require some form of coupling to the load.

**15.2** Discuss and analyze the discrepancy between the ideal trapezoidal waveforms of the brushless DC drive of Figure 17.4 and the actual waveforms that would result if winding resistance and practical transistor switching characteristics were taken into account. Explain how this leads to torque ripple in brushless DC motors.

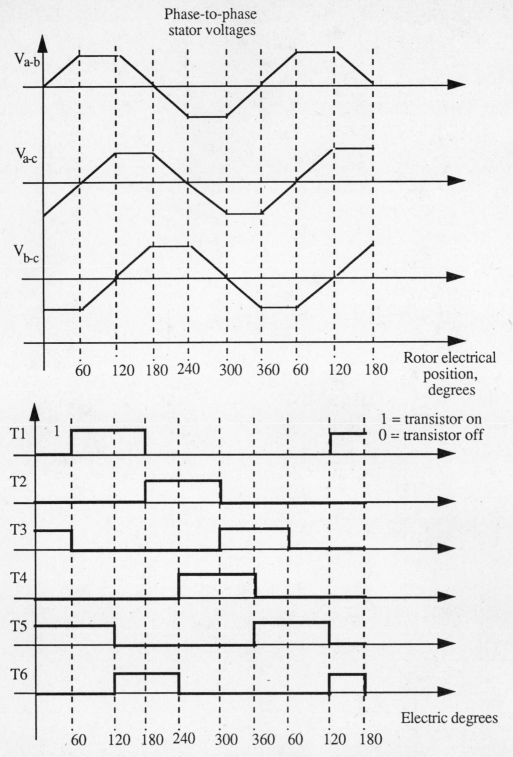

**Figure 17.4 Phase voltages and transistor (SCR) switching sequence for brushless DC motor drive of Figure 17.3**

**Solution:**

The impossibility to obtain exactly trapezoidal currents in practice, due to asymmetries in the resistance of the different windings and to the non-ideal transistor switching characteristics, results in an imperfect superposition of the waveforms. Thus, the net current is not a constant, but has a "ripple" component. Since the current is not constant, the torque produced by the motor cannot be constant; thus the "torque ripple" unavoidably present in these motors.

---

**15.3** The half step excitation scheme of a stepper motor is a combination of the single-phase and two-phase excitations. This scheme reduces the step angle by half. Derive the excitation sequence for a 3 phase motor and verify the result.

**Solution:**

The scheme is as follows:

| $i_a$ | $i_b$ | $i_c$ |
|-------|-------|-------|
| + | 0 | 0 |
| + | + | 0 |
| 0 | + | 0 |
| 0 | + | + |
| 0 | 0 | + |
| + | 0 | + |
| + | 0 | 0 |

---

**15.4** For a two-phase, two-pole rotor PM step motor, derive a switching sequence for a continuous rotation and draw the waveform of on-off periods of circuits in each of the two phases.

**Solution:**

The two-phase, two-pole permanent magnet step motor is shown below:

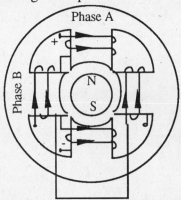

The switching sequence is as follows:

| Cycle | Phase A | Phase B | Position |
|-------|---------|---------|----------|
| + | 1 | 0 | 0 |
| | 0 | 1 | 90° |
| - | -1 | 0 | 180° |
| | 0 | -1 | 270° |
| + | 1 | 0 | 360° |

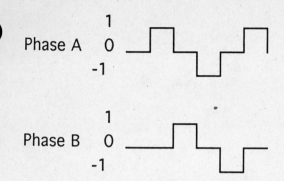

Phase A

Phase B

---

**15.5** A sine-cosine resolver is shown in Figure P16.3.

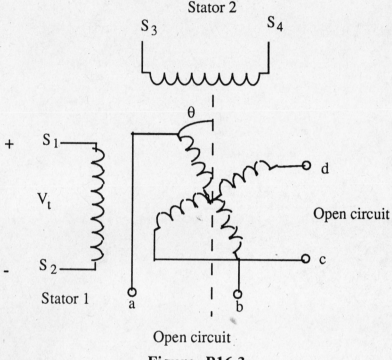

**Figure P16.3**

The stator winding $S_1S_2$ is supplied by excitation voltage $v_t = V_t \sin\omega t$. When $\theta = 0$, that is, the axis of winding ab is in alignment with the axis of $S_1S_2$ winding, the induced voltage in the rotor winding ab will be $E_{ab} = KV_t \sin\omega t$ and cd= 0, where K is the effective turns ratio of rotor winding to the stator winding. If the rotor is turned counter-clockwise for an angle of $\theta$, the induced voltage produced by forward field is $E^+_{ab} = \frac{1}{2}KV_t \sin(\omega t - \theta)$, the induced voltage produced by backward field is $E^-_{ab} = \frac{1}{2}V_t \sin(\omega t + \theta)$. Find the total voltage in winding cd. Verify that the magnitude of the $E_{cd}$ is the cosine function of the angle $\theta$.

**Solution:**

We have $E_{cd}^+ = \frac{1}{2}KV_t\cos(\omega t - \theta)$

$$E_{cd}^- = \frac{1}{2}KV_t\cos(\omega t + \theta)$$

$$E_{cd} = E_{cd}^+ + E_{cd}^- =$$

$$= \frac{1}{2}KV_t\cos(\omega t - \theta) + \frac{1}{2}KV_t\cos(\omega t + \theta)$$

$$= \frac{1}{2}KV_t\cos\theta\cos\omega t = |E_{cd}|\cos\omega t$$

where $|E_{cd}| = \frac{1}{2}KV_t\cos\theta$.

Therefore, the magnitude of the voltage $E_{cd}$ is proportional to the cosine of the angle $\theta$.

---